THE DELICATE ART OF BRUTE FORCE

Also by Paul J. Nahin

Oliver Heaviside (1988, 2002)
Time Machines (1993, 1999)
The Science of Radio (1996, 2001)
Time Travel (1997, 2011)
An Imaginary Tale (1998, 2007, 2010)
Duelling Idiots (2000, 2002)
When Least Is Best (2004, 2007, 2021)
Dr. Euler's Fabulous Formula (2006, 2011)
Chases and Escapes (2007, 2012)
Digital Dice (2008, 2013)
Mrs. Perkins's Electric Quilt (2009)
Number-Crunching (2011)
The Logician and the Engineer (2012, 2017)
Will You Be Alive Ten Years from Now? (2013)
Holy Sci-Fi! (2014)
Inside Interesting Integrals (2015, 2020)
In Praise of Simple Physics (2016, 2017)
Time Machine Tales (2017)
How to Fall Slower than Gravity (2018)
Transients for Electrical Engineers (2019)
Hot Molecules, Cold Electrons (2020, 2022)
In Pursuit of Zeta-3 (2021, 2023)
The Probability Integral (2023)
The Mathematical Radio (2024)

The Delicate Art of Brute Force

HOW TO COMPUTE IT WHEN YOU CAN'T SOLVE IT

PAUL J. NAHIN

PRINCETON UNIVERSITY PRESS
PRINCETON & OXFORD

Published by Princeton University Press
41 William Street, Princeton, New Jersey 08540
99 Banbury Road, Oxford OX2 6JX

press.princeton.edu

GPSR Authorized Representative: Easy Access System Europe - Mustamäe tee 50, 10621 Tallinn, Estonia, gpsr.requests@easproject.com

ISBN 978-0-691-26746-3
ISBN (e-book) 978-0-691-26752-4
Library of Congress Control Number: 2025946497

British Library Cataloging-in-Publication Data is available

Editorial: Diana Gillooly and Whitney Rauenhorst
Production Editorial: Kathleen Cioffi
Text Design: Wanda España
Jacket Design: Chris Ferrante
Production: Erin Suydam
Publicity: Matthew Taylor and Kate Farquhar-Thomson

This book has been composed in Arno Pro

Printed in the United States of America

10 9 8 7 6 5 4 3 2 1

IN MEMORY OF

Douglas George Nahin

1943–2024

A loving husband to his wife, Mary

A great father to his children, Yvette, Yvonne, and Douglas, Jr.

A wonderful brother

"Playing the game" was once the only way to study games of a random nature. The painting *The Dice Players* (1651) by French artist Georges de La Tour (1593–1652) shows that actually observing how the dice fall could capture the full attention of seventeenth-century gamblers. When de La Tour created his masterpiece of chiaroscuro, the mathematics of probability was just being developed, but then very quickly *theory* came to define the gold standard for studying games of chance (and remained so for centuries). Then, nearly 300 years after *The Dice Players,* the invention of the high-speed electronic digital computer created a dramatic (some even thought it to be a fantastic) new way to study probabilistic problems, as well as many other problems in which randomness is *not* a feature. Image courtesy of Wikimedia Commons and of the photographer and copyright holder of the image photograph, Thomas William Kirby.

CONTENTS

PRELUDE

The Birth of Electronic Computation

> It is naïve to expect that all mathematicians will be willing or even able to introduce computer-oriented materials in their courses. Nevertheless, numerical techniques must become an integral part of our undergraduate mathematics offerings.
>
> —J. E. MCKENNA, MARCH 1972[1]

THIS BOOK consists of a small number of master classes I have taught in the form of essays on the sort of problems high-speed electronic digital computers can "solve" that would otherwise be quite difficult (or even impossible) to handle analytically. Each problem uses the ability of modern computers to rapidly perform *massive* amounts of computation. I think it pretty obvious (because this book is limited, by practical considerations, to not exceeding a weight that can be carried by a reasonably fit person) that I cannot represent every conceivable type of problem here. However, I have attempted to include problems that typically confront applied mathematicians, engineers, and scientists. This focus is in contrast to a previous book (*Duelling Idiots*, Princeton 2002), in which I treated problems that are more whimsical than "serious," while the problems in this book are serious. I think of this text as being a "Goldilocks and porridge" book: that is, not too big and not too small, but just the right size to be satisfying without leaving you with an overstuffed feeling![2]

Before starting my academic career, I worked for over ten years in industry and military think tanks, and much of what you will see here actually came across my desk at one time or another. Often with a note from my then boss saying, in effect, "Think about this and see what you can do. I'd like to hear back from you by the end of the week." Sometimes that was the very next day!

I have selected the topics discussed using the following three criteria (and one more I will tell about in just a moment):

(1) Each arises from a well-defined physical process and is not just an abstract construction;
(2) The questions raised by each essay are easily understandable by a nonmathematician;
(3) The usual mathematical techniques taught to advanced high school and undergraduate physics and engineering students don't have any immediately obvious use.

The physical root of each topic gives a lot of nutritious food for thought. In general, computation, like any part of mathematics, is something to think about before doing it. And in each essay I foreground the thought processes that go into understanding a problem and then setting up a solution that a computer can produce. As a side benefit, I also introduce interesting and useful computational methods.

For instance, a particular example of applying criterion (3) is the use of random numbers in computer simulation studies. The invention of the high-speed computer powered the advancement of the theory of random numbers and the means for their fast generation by algorithmic (as opposed to physical) means. A number of the problems in this book illustrate how computers and random numbers are intimately tied together. It is somewhat ironic that computers are equally at home with *non*-random processes as well, and you will see that here, too!

Numerous books tackle the general topic of computers and number crunching (see, for example, my *Number-Crunching*, Princeton 2011), with most written for an assumed audience of fairly sophisticated readers (advanced college undergraduates and higher). The examples in these books are generally of an equally sophisticated nature (discussions of linear and dynamic programming are popular choices).[3] My goal here is to focus instead on the essentials that illuminate the art that lies at the heart of effective computation.

There is nothing in this book that attentive high school students who have taken an AP-calculus or AP-statistics class will find beyond them. This assumption lets me, for example, write (as I do in the first chapter) the symbol $\binom{n}{k}$ without explanation, with the expectation that a reader will instantly recognize it as denoting the binomial coefficient $\binom{n}{k}=\frac{n!}{k!(n-k)!}$, as well as be familiar with its physical interpretation as the number of different ways you can select (in any order) k objects from n distinguishable objects. I also assume some elementary probability knowledge at the level of knowing that if A and B are *independent* events, then $Prob(AB)=Prob(A)Prob(B)$. Beyond that, I offer some additional guidance in the Appendix which will (I hope!) address the most likely issues to cause difficulty. If your reaction to this is "This is old hat," *that's good*. You are just the sort of reader I had in mind as I wrote.

As for computer programming, my basic assumption is not much more than that of imagining my readers, as they open this book, will have an understanding of what a program is, and will have written some simple ones in a high-level language (this is, I think, actually a pretty weak assumption for today's high school AP students).

Okay, what about that fourth criterion (briefly mentioned in passing earlier) that I used to select the topics in this book? The whole point of this book is to show you, by detailed examples, how using a computer can (*sometimes*) succeed in getting an answer when analytical methods are not available (for example, you just don't know enough math). The problem I faced as I wrote is the obvious one of *how do I convince you that the computer answer is "correct"?* My response to that was to select problems that, besides looking plenty tough, also do have known analytical solutions. Each essay discusses both approaches, and I think you will be impressed by how well the computer solutions agree with theory. There are even situations where the computer solution not only is accurate but also gives additional insight and flexibility.

Indeed, the massive computational power of modern electronic computers has impressed just about all who are aware of just *how* huge that power is. For example, the video game console released in 2020 by Sony, the PlayStation®Five (PS5), performs its calculations at a rate just over 10 teraflops. That is, at ten *trillion*—and that is not a typo!—floating point operations per second. That blazing-fast performance is from a machine the size of a large shoebox,[4] using just 300 watts, that sells for under $500. This machine is available *not* by special access to a super-top-secret black-ops government lab based in the heavily guarded Area-51—that is, *not* a machine using advanced technology recovered from the remains of a crashed alien spacecraft—but rather a machine available over-the-counter at the video game shop in the local mall. Tens of millions of these machines (and of its competitor, the equally powerful Microsoft Xbox™) are in the bedrooms of teenagers worldwide (and in the offices of somewhat older gamers who, now and then, enjoy the cathartic release of a good first-person adventure/action game).

Today's supercomputers, operated by the US and some foreign governments, leave the PS5 and Xbox in the dust: such machines operate at speeds measured in *exaflops*. One exaflop is 10^{18} floating point operations per second! The coming of quantum computers suggests that even that astounding speed will eventually be eclipsed, but that's a tale for another book. We will be content, here, with the typical speed of a moderately priced laptop.

The ability—hour-after-day-after-week . . . —to tirelessly do elementary arithmetic operations at electronic speed has, alas, failed to impress *everybody*. Many (maybe not even a majority but still a lot of) "pure" mathematicians still cling to the Victorian belief that "real mathematics" needs no more than a brain, a piece of chalk, and a blackboard. When, in 1976, University of Illinois mathematicians Ken Appel (1932–2013) and Wolfgang Haken (1928–2022) used a computer to show the truth of a century-old conjecture on the coloring of planar maps,[5] a not uncommon response to their "proof" (an immense computer listing commonly said to resemble a telephone directory) was that of one mathematician who exclaimed, "I can't believe God would let such a beautiful theorem have such an ugly proof!"

That attitude exists despite the fact that such a superstar of mathematics as John von Neumann (1903–1957), at the Institute for Advanced Study (IAS) in Princeton, New Jersey (see Figure 1.1.1), enthusiastically supported using computers to attack complex problems in mathematical physics (such as long-range weather forecasting, and the propagation of chain-reaction neutrons in various nuclear bomb warhead geometries).

As World War II reached its end, the high-speed *electronic* computer came into existence in the form of the ENIAC (Electronic Numerical Integrator and Computer) at the Moore School of Electrical Engineering at the University of Pennsylvania. Computing machines before ENIAC (such as the Automatic Sequence Controlled Calculator [ASCC] at Harvard University) used electromechanical relay circuitry, which is both slow and vulnerable to random failure.[6] The ASCC had 3,500 relays and was said to sound like a roomful of clicking knitting needles when operating: the clicking of the ASCC (and of other relay computers) made a profound impression on all who heard it, and the noise remained a signature characteristic for all the *electronic* computers that came after it.

For example, in his famous story "The Last Question," science fiction writer Isaac Asimov describes the clicking of a huge, superadvanced computer called *Multivac*, despite the fact that its name is short for *multiple vacuum tubes*! That story appeared in 1956 more than a decade after the construction of the ENIAC (with its quiet vacuum-tube circuitry). Two years earlier, Asimov's science fiction colleague Fredric Brown had published the equally famous tale "Answer." In it an enormous computer (made by connecting together "all the monster computing machines of all the populated planets in the universe—ninety-six billion planets") is asked, "Is there a God?" To that came this instant response ("without the clicking of a single relay"): "Yes, *now* there is a God," along with a lightning bolt from a cloudless sky that melted all electrical circuits in-place, making them forever-after unalterable.

Science fiction writer Arthur C. Clarke's mid-1960s story "Dial F for Frankenstein" presented a similar cautionary view of the new digital age. Clarke imagined that all of the computerized telephone exchanges on Earth had been, at last (the story is set in the then near-future of 1975), connected via satellite. As one character explains, "Until today [our telephone networks] have been largely independent, autonomous. But now we've suddenly multiplied the connecting links, the networks have all merged together [like the neurons in a developing brain], and we've reached criticality." When another character asks what that means, the answer is chilling: "For want of a better word—consciousness." That is, Clarke's tale envisioned the awakening of a self-aware computer long before the *Terminator* films introduced Skynet to movie audiences. Since Clarke's writing, satellites *have* connected all the world's telephone networks, and nothing untold has happened. At least, I don't think so. Of course, the really smart thing for a self-aware computer to do would be to remain quiet about who is *really* running the show and so, maybe, . . . ?

Well, perhaps there is *some* reasonable concern about that possibility, but in this book I limit myself to endorsing a somewhat less terrifying view of computers!

ENIAC's electronic circuitry was more than two thousand times faster than the ASCC's relays. In the years following

ENIAC's birth, ever-more sophisticated electronic machines were built, with one of the more famous being the MANIAC (Mathematical Analyzer, Numerical Integrator and Computer) at the Los Alamos Scientific (later, National) Laboratory, New Mexico, where the first atomic bombs were made. The MANIAC, able to do 11,000 additions per second, helped make important advances in answering questions concerning atomic physics, and physicists embraced it with great enthusiasm; in particular, Nobel Prize winner Enrico Fermi (1901–1954).[7]

The development of ENIAC opened a vast new world of possibilities for physicists and engineers. Suddenly, problems that had been simply too complex, difficult, or messy for analytical treatment could now be defeated by the application of pure, brute force; that is, by the ability of electronic machines to perform an enormous number of simple operations in just minutes. As Lord Kelvin said in the nineteenth century, about a mechanical analog computer he had built for tide predictions, his machine "substituted brass for brains."[8] Compared to modern machines, or even to machines that would be made less than ten years later, ENIAC was pretty elementary—but it was good enough to give analysts a strong hint as to what was to come. Years earlier analysts already sensed that physics had reached such a level of sophistication that *something new* was required to handle the torrent of math regularly being encountered. In the last year of the 1920s, English theoretical physicist Paul Dirac (1902–1984)—just four years before he received the 1933 Nobel Prize in physics—opened a paper[9] with this bold claim: "The general theory of quantum mechanics is now almost complete" and so "the underlying physical laws necessary for the mathematical theory of a large part of physics and the whole of chemistry are thus completely known, *and the difficulty is only that the exact application of these laws leads to equations much too complicated to be soluble* [my emphasis]."

And then came ENIAC (followed by MANIAC, in parallel with von Neumann's IAS computer at Princeton), the beginning of what now seems to be an endless stream of ever-more advanced high-speed electronic computers, as answers to Dirac's "difficulty"—answers that even his great genius could not have foreseen in 1929.

Now, before we get started in earnest, here is a quick, elementary example of how the development of electronic computers introduced a paradigm shift in analysts' approach to grubby math problems. To bring new staff members quickly up-to-speed on the atomic bomb project at Los Alamos, in 1943 physicist Robert Serber (1909–1997) gave a series of elementary lectures on the physics of the bomb.[10] At one point the equation $xcos(x) = (1 - a)\ sin(x)$ appeared, where a is some value in the interval 0 to 1. It was important to solve this equation for x as a function of a. The obvious thing to do is to plot, for a given a, each side of the equation and then observe where the two curves cross. This was done by hand, using a good set of math tables, a sharp pencil, and a sheet of graph paper. A new plot was needed for each new value of a. Can you imagine how boring this must have been?

In Figure P1, I've shown how my "modern" computer (a four-year-old laptop) makes short work of this, where the light curve is $xcos(x)$ and the dark curve is $(1 - a)\ sin(x)$ for the case of $a = \frac{1}{2}$. You can see from the figure that there are an infinite number of crossings of the two curves, with the first one agreeing with Serber's value in the *Primer* of $x \approx 1.17$. If a bomb analyst wanted to plot a new curve showing how the solution x varies as a function of a (and so he needed, perhaps, 100 solutions), you can appreciate how easy that would be to do using a computer, and also how *miserable* it would be to do by hand.

So far, I have concentrated on the hardware side of the electronic computer revolution. Of equal importance is the software side that allows humans to "tell" a computer what needs to be done and how to do it. With MANIAC that was done by using what computer scientists call *machine language* coding—that is, the writing out, in great detail, a specific sequence of elementary steps to be performed using the fundamental commands built

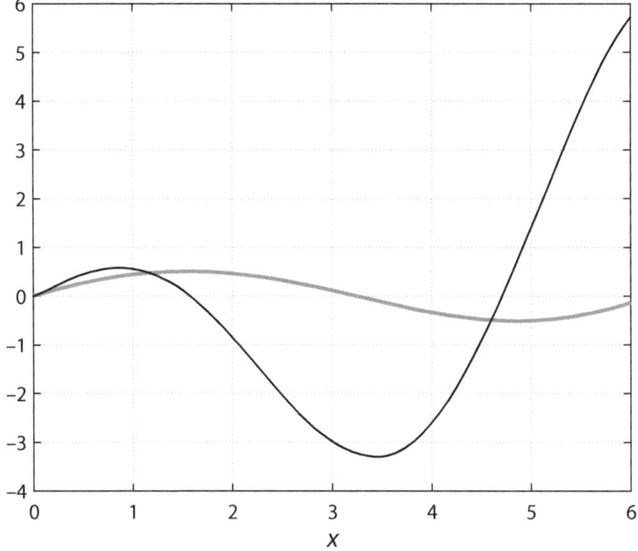

FIGURE P1. The Los Alamos equation for $a = 1/2$.

into the computer's circuitry by its designers. This work is very demanding, time-consuming, and downright tedious (I write that from personal experience). The development of special programs called *assemblers, interpreters,* and *compilers* transferred most of that work to the computer itself, which made writing and debugging programs vastly easier than it was for MANIAC (see Anderson's paper in note 7 for what machine language looked like[11]). With MANIAC, if you had a number x and wanted to compute $cos(x)$, you had to either code the power series expansion for the cosine function, evaluate term by term, and then sum, or store a table of precomputed values of the cosine function, do a table lookup, and then (perhaps) interpolate. Today you simply write $cos(x)$ and the compiler translates that into the machine language the computer hardware "understands."

All the codes created in support of this book have been written in MATLAB, a widely used scientific programming language popular among not only engineers but also many physicists and mathematicians. However, you do *not* need to know MATLAB to read this book, as our central interest is not in computer programming but rather in developing *algorithms* for solving our problems. Still, just to show what a MATLAB code looks like, what follows is the code that generated Figure P1, and I think you will find it almost self-explanatory.

```
%warmup.m
a=0.5;x=linspace(0,6,1000);
left=x.*cos(x);
right=(1-a)*sin(x);
plot(x,left,'-k')
hold on
plot(x,right,'.k')
xlabel('x')
grid on
```

MATLAB's instruction set is absolutely *enormous*, with numerous specialized commands for doing an equally enormous number of exotic operations. For the most part, however, in this book all the codes shown are not particularly deep, and you should find translating to any other high-level programming language, like Python, easy to do, as assignment statements, and *for*, *if*, and *while* loops, are about as sophisticated as I get. Many authors try to avoid the specificity of using a particular language in which to write code by using a *pseudocode*, a made-up, "self-evident" (so it is hoped) language. My elementary use of MATLAB is, I assert, so close to being pseudocode as *to be* pseudocode. The central core of each code is, in any case, not the code itself, but rather the *algorithm* that the code implements.

Okay, the above code is a *warm-up*. Let's now see what a computer can do with some *really nasty* math problems.[12] When I say we are going to be discussing math problems, I mean *applied* math problems (with one exception, in chapter 6). Our discussions here will be of no interest to ultra-pure mathematicians like the late G. H. Hardy.[13]

We will start with an easy problem (basically, one of pure number crunching) and get progressively more complicated as we move into the book. These problems will illuminate the motivation behind the following words, written by MANIAC's lead designer fifteen years after the quotation that opens this Prelude:

> It is interesting to look back over two score years and note the emergence . . . of experimental mathematics, a natural consequence of the electronic computer. . . . At long last, mathematics [has] achieved a certain parity—the twofold aspect of experiment and theory—that all other sciences enjoy.[14]

Metropolis had, in a way, been anticipated in this sentiment by his counterpart at Princeton, Julian Bigelow (1913–2003), chief engineer on the IAS Electronic Computer Project. In 1976 Bigelow declared that, even as he and his team worked, they had fully understood what they were creating. Because of MANIAC and the IAS machine "A tidal wave of computational power was about to break and inundate everything in science and much elsewhere, and things would never be the same."[15]

Across the Atlantic came the equally awed words of British mathematical physicist Douglas Hartree (1897–1958), who had spent time in America learning how to program the ENIAC.[16] In a November 1946 newspaper interview about a new computer called ACE (Automatic Computing Engine) being developed at London's National Physical Laboratory, he declared, "The implications of the machine are so vast that we cannot conceive how they will affect our civilization. Here you have something which is making one field of human activity 1,000 times faster. In the field of transportation, the equivalent to ACE would be the ability to travel from London to Cambridge [about 58 miles] in five seconds as a regular thing. It is almost unimaginable."

For some, the words of Metropolis, Bigelow, and Hartree produce an ominous sense of dread[17] (recall the cautionary science fiction tales of Brown and Clarke). For example, theoretical

FIGURE P2. The MANIAC-I electronic (vacuum tube) computer (1952). Image courtesy of the Los Alamos National Laboratory.

physicist (Harvard PhD)–turned–literary essayist Jeremy Bernstein (1929–) once approached von Neumann about this very issue. As he wrote in one of his many elegant *New Yorker* magazine essays (as a staff writer), "As a graduate student, I once asked the late Professor John von Neumann whether he thought computers would replace mathematicians. His answer was, 'Sonny, don't worry about it.'"[18]

Was von Neumann's seeming unconcern warranted? I don't know, but certainly others remain convinced that such a cavalier dismissal may deserve a revisit. As the opening line of a review of a recent, impressively imaginative "biography" of von Neumann declares, "If the most dangerous invention to emerge from World War II was the atomic bomb, the computer now seems to be running a close second, thanks to recent developments in artificial intelligence."[19] Bigelow and Metropolis never expressed similar concerns, but they did live long enough to see how well they had predicted the future of computation. (Hartree died too young to really have a chance to see what was soon to come.) Indeed, that future arrived with what a mystic might say was a speed far faster than that of time itself. As a contemporary pointed out to me, after reading an early draft of this Prelude, "We were lucky. We were both born before ENIAC and we are both still alive today. The entirety of the electronic digital age."[20]

THE DELICATE ART OF BRUTE FORCE

1

Computers That Play Games

1.1 The IAS and MANIAC Machines

The first two sections of this chapter set the historical stage for the computational problem we will be tackling. If you are too curious to wait, you can turn to section 1.3 right now (but I think you will find the history fascinating as well).

One of the early fans of electronic computers, whose excitement about ENIAC was at least as great as Fermi's, was von Neumann, and he wanted to build an even more powerful computer at Princeton. The IAS Electronic Computer Project—which ran, essentially, in parallel with the development of MANIAC-I from 1948 onward—almost immediately encountered severe criticism from von Neumann's IAS mathematician colleagues (recall my comments in the Prelude about pure mathematicians and computers). They felt that the *building* of something would be an exercise in "mere engineering" and, so, unworthy of the Institute. In the end, von Neumann prevailed, but low-level grumbling among many of the IAS mathematicians continued for quite a while.

British novelist Robert Harris perfectly captured, in his 2020 novel *V2* (Random House), how the IAS mathematicians viewed von Neumann's computer people. On the first page of his novel, Harris describes a group of men gathered in late 1944 at a V2 rocket launch site in Holland, just before one of these devastating

FIGURE 1.1.1. John von Neumann and the IAS machine, 1951. Image courtesy of the Shelby White and Leon Levy Archive Center, Institute for Advanced Study (Princeton, N.J.), Alan W. Richards, photographer.

weapons is to be sent on its way to London. All are clearly military men, with one exception.

> He was the only one not in uniform. His pre-war dark blue suit with its row of pens in the breast pocket, along with his worn plaid tie, proclaimed him a civilian—a maths teacher, you might have said if you had been asked to guess his profession, or a young university lecturer in one of the sciences. Only if you noticed the oil beneath his bitten finger-nails might you have thought: ah-yes—an engineer.

For (many) pure mathematicians, chalk dust on the nose is the proud signature of academic purity, but oil under the fingernails identifies someone who simply "builds things." Whether it is rockets or computers is a distinction that hardly matters to the purist.

The IAS machine and MANIAC-I (each using electronic circuitry involving *thousands* of vacuum tubes, as did ENIAC) both went operational in early 1952. (In later years, advanced versions of the original Los Alamos machine, named MANIAC-II and MANIAC-III, were built, using ever-newer technology. MANIAC-III, for example, used transistor circuitry.) All these machines were eventually used in support of developing thermonuclear fusion bombs (the H-bomb), but the Los Alamos MANIAC-I had some interesting adventures in addition to that of simply designing bigger bombs.[1]

1.2 MANIAC and Chess

MANIAC-I was used in a wide variety of pioneering studies, some of which are briefly described in Anderson's essay (note 7 in the Prelude).[2] I think the study that best illustrates an early appreciation of how ever-increasing computer power was going to change the world was the programming of MANIAC-I in the mid-1950s to play chess. The idea of programming a computer to play chess had, in fact, been around before MANIAC-I. The well-known Bell Telephone Laboratories electrical engineer and mathematician Claude Shannon (1916–2001) had, a few years earlier, published a long, detailed essay on how that might be done.[3] As a purely theoretical paper, it made no attempt to *build* a chess computer. A less ambitious attempt to *construct* a game computer was made, however, just a couple of years after Shannon's paper appeared, when IBM computer scientist Arthur Samuel (1901–1990) programmed an IBM 701 to play a fairly decent game of checkers. In 2007 Samuel's goal of a perfect checkers computer was at last achieved, and today it is impossible for a human to beat the computer (a draw is the best a human can hope for, and then only by not making even a single mistake).[4] These early efforts mark, I believe, the birth of artificial intelligence.

Chess is a vastly more complex game than is checkers, and so it presented a far greater challenge for the MANIAC programmers.[5] Shannon's paper gives a simple but dramatic way to see this. In chess a full move consists of one player moving a piece *followed* by the other player moving a piece; that is, each player performs a *half*-move in this exchange. Shannon argued that, on average, each player typically has something like 30 possible legal options available for her half-turn, giving $30^2 \approx 1{,}000 = 10^3$ possibilities for each full move. Since the typical chess game lasts for about 40 full moves, there are a total of $(10^3)^{40} = 10^{120}$ chains of full moves, that is, 10^{120} possible chess games. Even if a computer could follow each chain from its start to its finish, at the rate of one million chains per second, it would take (said Shannon) 10^{90} years to consider them all.[6]

Shannon's paper points out that in chess there is no random element (other than the initial decision of who plays White and so goes first). In addition, both players have complete access to all information (there are no hidden variables) at every moment. Thus, as shown in the classic book[7] on game theory (coauthored by von Neumann), there exists a strategy **S** such that, given any initial position of the pieces, there are just three possibilities:

(1) It is a won position for White. That is, White can force a win no matter how Black plays;
(2) It is a drawn position. That is, both White and Black can at least force a draw no matter what the other player does. If both players don't deviate from **S** (whatever it is), the game *will be* a draw;
(3) It is a won position for Black.[8] That is, Black can force a win no matter how White plays.

The problem for would-be chess computer programmers is that while the optimal **S** exists, nobody has the slightest idea what it is!

And maybe that is not such a terrible thing. If **S** *were* known, then Shannon imagines the following gloomy (for chess lovers) scenario:

> The unlimited intellect assumed in the theory of games . . . never makes a mistake. . . . A game between such mental giants, Mr. A and Mr. B, would proceed as follows. They sit down at the chessboard, [decide who plays White], and then survey the pieces for a moment, Then either:—
>
> (1) Mr. A says, "I resign," or
> (2) Mr. B says, "I resign," or
> (3) Mr. A says, "I offer a draw," and Mr. B replies, "I accept."

What great fun *that* would be, right? Almost certainly *not*!

Shannon spends the rest of his paper discussing several ways to develop an *effective* strategy after having observed, "It is clear that the problem is not that of designing a machine to play perfect chess [as Schaeffer did for checkers, a task Shannon incorrectly declared to be "quite impractical"] nor one which merely plays legal chess (which is trivial). We would like [the computer] to play a skillful game, perhaps comparable to that of a good human player."

Shannon lived just long enough to see (as I will explain in a moment) this ambitious goal exceeded *by far*.

Not having **S** available meant that the Los Alamos programmers had to develop a heuristic algorithm that did the best it could without attempting an exhaustive, brute-force examination of all possible games. Such an algorithm was developed for MANIAC-I by a small group of Los Alamos scientists headed by famous mathematician Stanisław Ulam (1909–1984), the same Ulam mentioned in note 2. MANIAC-I was in great demand by many of the Los Alamos scientists, each with their own special project, and I suspect Ulam's primary role was using his very senior status to simply gain access to the machine. The actual programming was, I believe, done by two young mathematicians, Paul Stein (1924–1990) and Mark Wells (1929–2018).

Stein and Ulam published a quite interesting report on MANIAC-I's introduction to what has become known as "Los Alamos chess," or "6 × 6 chess," or, most interesting of all, "anti-clerical chess."[9] The names come from the fact that, for MANIAC-I's computational limitations to be able to look even just two moves ahead (a good chess player typically looks six or even more moves ahead), the game was reduced in complexity from an 8-by-8 board to a 6-by-6 board by removing the bishops. Even so, it took MANIAC-I about 12 minutes to make a decision for each of its moves.[10] Stein and Ulam discussed, in great detail, three games played by MANIAC-I (against itself, against a strong human player, and against a weak human player). Their general feeling was that the machine's performance was that of a human "who has average aptitude for the game and experience amounting to 20 or so full games played."

Chess players are rated on a numerical scale that has beginners scored from 100 or so to perhaps the 900s. Even better players have ratings from over 1,000 to almost 2,000, tournament players (up to so-called grand masters) are at about 2,000 to 2,400, and world champions are at somewhere like 2,900. MANIAC-I was probably playing with a rating of less than 500.

How things have changed since Stein and Ulam wrote. Unlike checkers, chess has not been solved. That is, while the perfect strategy for chess exists, it is still unknown. However, the *heuristic* strategies for chess are now so powerful that chess computers are already essentially unbeatable. The strongest chess computer code (called *Stockfish*) prior to 2017 has a rating of about 3,500; it is so strong that the current human world champion refuses to play against it. *Stockfish* is almost certain to win any individual game it plays against any human, and is *virtually certain* to win a match involving multiple games.[11] The editors of *Chess Review* magazine wrote a "cry from the heart" *Afterword* to the Stein/Ulam article (note 9), in which they declared, "As devotees to chess, and so possibly biased, we feel that the game cannot be reduced to any mathematical formula—not even as complicated

FIGURE 1.2.1. Paul Stein (on the left) and Nicholas Metropolis (lead designer of MANIAC-I) playing Los Alamos chess against the MANIAC-I (which is behind them). Image courtesy of The Los Alamos National Laboratory.

and extensive as Einstein's tensor equations [of general relativity]." That is a very high bar to clear, and the editors might well be right, even today, nearly seventy years later, but in fact it no longer really matters. Even without knowledge of the perfect strategy **S**, reality has already reached the state of "game over" for even the very best human chess players who dare to challenge the computer.

1.3 The Probability of a Tied Match

So now, at last, we come to the computer problem for this opening chapter, a problem suggested by the following words from the opening of a paper by a DePaul University mathematician: "[The IBM computer code] *Deep Blue* and Gary Kasparov recently played [a six-game match]. My son, Andrew, thought that was a bad idea to have the number of games be even, because this would make the probability of a tie for the match too high. That seemed like pretty sound intuition to me. What follows is an analysis of a fairly realistic model of match play between approximately equal players. *In this model Andrew's intuition fails* [my emphasis]."[12]

The occurrence of a tie in a high-profile competition is not a good outcome for the promoters of the event. What excites people, after all, is the emergence of a *winner* (and, of course, a *loser*). A tied match doesn't leave people with a satisfied feeling—so, what *is* the probability of a tied match? Before Ash's analysis, the general belief was that of Andrew's: a value unacceptably "too high." Ash showed that belief isn't correct.

We can get some preliminary insight into how Andrew's intuition was faulty by considering the special case of equally matched players who *never* draw a game (any particular game *always* results in a win for one or the other of the two players). The only way a match of N such games can end in a tie is if each player wins $\frac{1}{2}N$ games (remember, N is even, as argued in note 11). This happens with probability

$$P_{tie} = \binom{N}{\frac{1}{2}N}\left(\frac{1}{2}\right)^{\frac{1}{2}N}\left(\frac{1}{2}\right)^{\frac{1}{2}N} = \frac{N!}{\left\{\left(\frac{1}{2}N\right)!\right\}^2}\frac{1}{2^N}. \qquad (1.3.1)$$

Both the numerator and the denominator in this last expression blow up as $N \to \infty$, and it may not be immediately obvious by inspection what P_{tie} does as $N \to \infty$. Numerical computation answers that for us, and Figure 1.3.1 is a plot of (1.3.1) as N varies through the even values from 2 to 200. The plot shows that P_{tie} decreases with increasing N (and so, for this particular case, Andrew's intuition does indeed fail).[13] An obvious follow-up question immediately presents itself: what is the nature of the inverse relationship between P_{tie} and N? That is, *how* does P_{tie} decrease with increasing N?

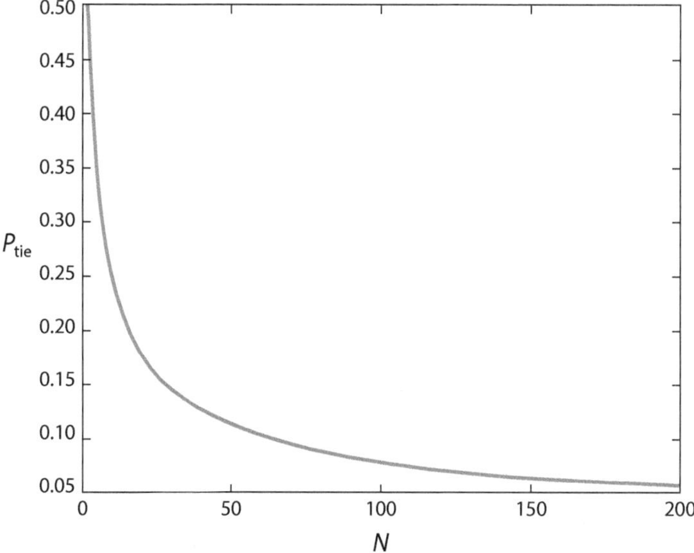

FIGURE 1.3.1. Probability of a tied match after N games between equal strength players who never draw a game.

To analytically explore what happens as we let N become "large," the tool to use is Stirling's asymptotic approximation[14] to $m!$, that is,

$$m! \sim \sqrt{2\pi m}\; m^m e^{-m} \text{ as } m \to \infty. \tag{1.3.2}$$

Thus, for large N we have

$$P_{tie} = \frac{\sqrt{2\pi N} N^N e^{-N}}{\left(\sqrt{2\pi \frac{N}{2}} \left(\frac{N}{2}\right)^{N/2} e^{-N/2}\right)^2 2^N}$$

$$= \frac{\sqrt{2\pi N} N^N e^{-N}}{2\pi \frac{N}{2} \left(\frac{N}{2}\right)^N e^{-N} 2^N}$$

$$= \frac{\sqrt{2\pi N} N^N}{2\pi \frac{N}{2} \frac{N^N}{2^N} 2^N} = \frac{2}{\sqrt{2\pi N}}.$$

or

$$P_{tie} = \sqrt{\frac{2}{\pi}} \frac{1}{\sqrt{N}} = \frac{0.7979}{\sqrt{N}} \tag{1.3.3}$$

which tells us that $\lim_{N\to\infty} P_{tie} = 0$. For $N = 100$, for example, $P_{tie} \approx 0.08$. It is interesting to note that for Stirling's approximation to be a "good" one, N doesn't actually have to be all that large. If $N = 6$, for example, then from (1.3.3) P_{tie} is approximately

$$\sqrt{\frac{2}{\pi}} \frac{1}{\sqrt{6}} = \frac{1}{\sqrt{3\pi}} = 0.3257$$

while the exact value is (from (1.3.1))

$$\frac{6!}{(3!)^2} \frac{1}{2^6} = \frac{5}{16} = 0.3125.$$

There is an elegant way to use a computer to show that the inverse square-root behavior is actually a pretty good approximation even for very small N. The hint (for large N) given us via Stirling, that $P_{tie} = \frac{k}{\sqrt{N}}$ where k is some constant, means that

$$\begin{aligned} ln(P_{tie}) &= ln\left(\frac{k}{\sqrt{N}}\right) = ln(k) + ln\left(\frac{1}{\sqrt{N}}\right) = ln(k) + ln(N^{-1/2}) \\ &= ln(k) - \frac{1}{2} ln(N). \end{aligned}$$

That is, if P_{tie} varies inversely as the square root of N, then $ln(P_{tie})$ varies *linearly* with $ln(N)$, and so a plot of $ln(P_{tie})$ versus $ln(N)$ will be a *straight line with negative slope* (this is *independent* of the value of k, which simply introduces a vertical shift). MATLAB can generate such a plot (called a *log-log plot* with each axis scaled logarithmically) with its wonderful *loglog* plotting command. Figure 1.3.2 shows the result: the solid line is a plot of the numerical values of P_{tie} calculated from (1.3.1), along with a plot of the reference line $\frac{1}{\sqrt{N}}$ (the dashed line). The two plots are, to the eye, parallel (equal slopes) *right from the starting value of* $N = 2$.

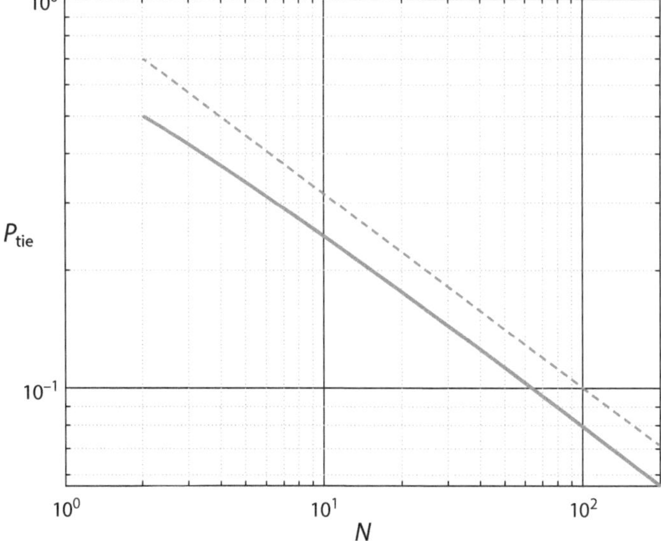

FIGURE 1.3.2. The loglog version of Figure 1.3.1. P_{tie} is the solid line calculated from (1.3.1), while the dashed line is the behavior of $1/\sqrt{N}$.

Now, it is not very realistic to assume our two players never draw any individual games. Indeed, the specific problem treated by Professor Ash was that of equal strength players *who do*, with probability q, draw any particular individual game. Specifically, Ash assumed each player has probability $p=\frac{1}{3}$ of winning an individual game, and so the probability of drawing a game is $q=1-p-p=\frac{1}{3}$. *For these particular numbers,* Ash showed (using sophisticated probability arguments that are a bit beyond the level of this book) that

$$\lim_{N\to\infty} P_{tie} = \sqrt{\frac{3}{4\pi}}\,\frac{1}{\sqrt{N}} = \frac{0.4886}{\sqrt{N}}.$$

Again, we see $\frac{1}{\sqrt{N}}$ behavior for P_{tie}. For $N=100$, for example, $P_{tie}\approx 0.05$, significantly *less* than P_{tie} for equal strength players who never draw a game. I personally find this nonintuitive.

With a computer, we can extend the study of tied *matches* between equal strength players who play tied *games*, to the even more realistic case of players with *unequal* strengths. For that situation, Professor Ash merely says that it might be an interesting thing to do and states (without supporting analysis), "the probability of a [tied match] decreases *exponentially* [my emphasis] as N increases." Is that correct? With a computer, we can explore this experimentally.

To formulate this generalization for computer study, let's write p for the probability player A wins a given game and, as before, q for the probability a game ends in a tie. That leaves probability $1-p-q$ for the probability player B wins a given game. Now, the only way a match of N games (N even) can end in a tie is if there have been d drawn games where d is even ($d=0$ or 2 or 4 or . . . N), leaving $N-d$ games (which is clearly even) to be split evenly between A and B. Since there are $\binom{N}{d}$ ways to select the d games that are draws, and since there are $\begin{pmatrix} N-d \\ \frac{1}{2}(N-d) \end{pmatrix}$ ways to select the $\frac{1}{2}(N-d)$ games that A wins (alternatively, the games that B wins), we arrive at the perhaps fearsome-looking

$$P_{tie} = \sum_{d=0,2,4,\ldots,N} q^d p^{\frac{1}{2}(N-d)} (1-p-q)^{\frac{1}{2}(N-d)} \begin{pmatrix} N \\ d \end{pmatrix} \begin{pmatrix} N-d \\ \frac{1}{2}(N-d) \end{pmatrix}. \tag{1.3.4}$$

It would not be very hard to convince you that computing by hand P_{tie} from (1.3.4), using numerous values for p, q, and N, is a task that would break the spirit of even the most ardent lover of arithmetic. For a modern electronic-speed, number-crunching computer, however, it is all duck soup.

To test Professor Ash's statement of exponential behavior for P_{tie}, let's *assume* the simplest possible form of

$$P_{tie} = ke^{-\alpha N} \tag{1.3.5}$$

where k and α are positive constants. Then,

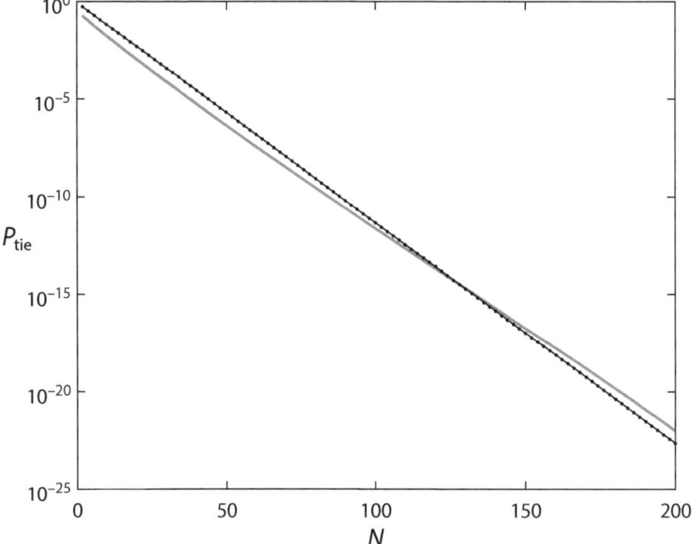

FIGURE 1.3.3. Semi-log plot of ln(P_{tie}) versus N from (1.3.4) (solid line), and the *assumed* exponential expression in (1.3.5) for $\alpha = 0.26$ (dotted line).

$$ln(P_{tie}) = ln(ke^{-\alpha N}) = ln(k) + ln(e^{-\alpha N})$$

or,

$$ln(P_{tie}) = ln(k) - \alpha N. \qquad (1.3.6)$$

That is, $ln(P_{tie})$ varies linearly with N. So, if we make a so-called *semi*-log plot of $ln(P_{tie})$ versus N (using MATLAB's convenient *semilogy* plotting command) to logarithmically scale the vertical axis alone (*semilogx* would logarithmically scale the horizontal axis alone), we should see a straight line with negative slope.

Figure 1.3.3, for example, is a semi-log plot of $ln(P_{tie})$ as N varies from 2 to 200 (the solid line) for the case of very *unequal* strength players with individual probabilities 0.6 and 0.1 of winning a game, and so a probability of 0.3 for drawing a game. The plot is, indeed, a straight line, and as a reference the figure also shows (the dotted line) what our *assumed* exponential behavior in (1.3.5) looks like for $\alpha = 0.26$ (a value found by simply running the computer code evaluating (1.3.4) numerous times, with various values of α, to *experimentally* see what works). As intuitively expected, P_{tie}, in the case where one player is significantly stronger than the other, is *dramatically* reduced from what it was for the case of equal strength players.

As far as I know, the analysis of tied matches today is pretty much as Professor Ash left it, and so this is a natural place to stop and move on to chapter 2.

2

A Gambling Problem (the Monte Carlo Concept)

2.1 Tossing Dice

Long before the invention of the high-speed electronic digital computer, scientists tackled—with some success—problems that had defied usual analytical methods by using, instead, the technique of *simulating the physics* of the problem. Those early problems were drawn from a broad spectrum of quite different physical processes, but they all shared one common feature: in some way or other they involved *randomness*. (This is a good time for you to look at the Appendix.)

Mathematicians will almost certainly be familiar with two particular such problems that are classics of their genre: Buffon's needle tossing experiment to estimate the value of pi, and the St. Petersburg paradox, each dating from the mid-eighteenth century.[1] Each is discussed in any good book on probability. The earliest *physics* problem attacked by simulation was, I believe, one Lord Kelvin discussed at the beginning of the twentieth century, in a study of the thermodynamics of a gas (see Kelvin's famous paper—"Nineteenth Century Clouds Over the Dynamical Theory of Heat and Light," *Philosophical Magazine*, July 1901, pp. 1–40—that addresses puzzles that couldn't be explained until the development of quantum mechanics). I say a few words in the Appendix about how Kelvin implemented the random part of his simulation.

In this chapter we will consider a problem in gambling (take another look at the frontispiece) that is very easy to understand, but is still sufficiently involved that I don't think you will be able to make much *analytical* headway with it without the aid of some probabilistic thinking. Then, after we have found the answer to the problem, I will show how a computer can make impressively short work of the problem *even if you know essentially nothing about probability*. So, first, the theoretical solution.

To give the problem a "personal touch," suppose you are the staff mathematician at a Las Vegas casino (I have no idea if such a job exists, but it *does* sound like fun, don't you think!), and your boss has asked you for a report on the following game. Imagine you have two fair dice that you toss (as a pair) six times. After each double toss you record the sum of the two faces that show. The possible sums are, of course, the integers 2 to 12. Our question is this: What is the probability the first sum is different from any of the next five sums? This might well have been the game being played by the frontispiece gamblers, who would have been greatly interested in the answer to help them decide how to bet. And, of course, *you* are greatly interested, as well, because your boss is waiting for an answer!

2.2 Theoretical Solution

To start, suppose the first sum is 2, which can occur in just one way (each of the dice shows a 1). Since there are $6 \times 6 = 36$ ways two dice *could* fall, the sum of 2 happens with probability $\frac{1}{36}$. The remaining 35 ways the two dice could fall on each double-toss results (of course) in sums other than 2, and so the probability of a sum other than 2 is $\frac{35}{36}$. Thus, given that the first sum is 2, the probability none of the next five double tosses give a sum matching the first sum is (assuming the double tosses are independent and so we can multiply probabilities) $\left(\frac{1}{36}\right)\left(\frac{35}{36}\right)^5$.

With similar reasoning (using the numbers in Table 2.2.1 that shows all the ways the sums 2 to 12 can occur), we get the total probability that the first sum does not match any of the next five sums to be

$$\left(\frac{1}{36}\right)\left(\frac{35}{36}\right)^5+\left(\frac{2}{36}\right)\left(\frac{34}{36}\right)^5+\left(\frac{3}{36}\right)\left(\frac{33}{36}\right)^5+\left(\frac{4}{36}\right)\left(\frac{32}{36}\right)^5$$
$$+\left(\frac{5}{36}\right)\left(\frac{31}{36}\right)^5+\left(\frac{6}{36}\right)\left(\frac{30}{36}\right)^5+\left(\frac{5}{36}\right)\left(\frac{31}{36}\right)^5+\left(\frac{4}{36}\right)\left(\frac{32}{36}\right)^5$$
$$+\left(\frac{3}{36}\right)\left(\frac{33}{36}\right)^5+\left(\frac{2}{36}\right)\left(\frac{34}{36}\right)^5+\left(\frac{1}{36}\right)\left(\frac{35}{36}\right)^5$$
$$=2\left(\frac{1}{36}\right)\left(\frac{35}{36}\right)^5+2\left(\frac{2}{36}\right)\left(\frac{34}{36}\right)^5+2\left(\frac{3}{36}\right)\left(\frac{33}{36}\right)^5$$
$$+2\left(\frac{4}{36}\right)\left(\frac{32}{36}\right)^5+2\left(\frac{5}{36}\right)\left(\frac{31}{36}\right)^5+\left(\frac{6}{36}\right)\left(\frac{30}{36}\right)^5$$
$$=\frac{2(35)^5+4(34)^5+6(33)^5+8(32)^5+10(31)^5+6(30)^5}{(36)^6}$$
$$=\frac{(105{,}043{,}750)+(181{,}741{,}696)+(234{,}812{,}358)+(268{,}435{,}456)+(286{,}291{,}510)+(145{,}800{,}00)}{2{,}176{,}782{,}336}$$
$$=\frac{1{,}222{,}124{,}770}{2{,}176{,}782{,}336}=\frac{611{,}062{,}385}{1{,}088{,}391{,}168}.$$

If you apply the Euclidean algorithm[2] to this last fraction, the result is the greatest common divisor of the numerator and the denominator is 1, and so the fraction has been reduced as far as possible. In any case, the exact probability we are after is 0.561436369 . . . . That is, more often than not there will *not* be a match of any of the last five sums with the first sum, a result that seems (for many people) a bit surprising.

TABLE 2.2.1. All Possible Results of Double Tossing Two Dice and Recording Their Sum

Sum	# of ways	How the sum can occur
2	1	(1,1)
3	2	(1,2), (2,1)
4	3	(1,3), (3,1), (2,2)
5	4	(1,4), (4,1), (2,3), (3,2)
6	5	(1,5), (5,1), (4,2), (2,4), (3,3)
7	6	(1,6), (6,1), (2,5), (5,2), (3,4), (4,3)
8	5	(6,2), (2,6), (3,5), (5,3), (4,4)
9	4	(6,3), (3,6), (4,5), (5,4)
10	3	(6,4), (4,6), (5,5)
11	2	(5,6), (6,5)
12	1	(6,6)

2.3 Computer Solution

Okay, suppose you do not have any probability knowledge to fall back on, other than being aware of the uniform random number generator available in any good scientific programming language (see the Appendix for a discussion of MATLAB's generator that, each time it is invoked, returns a number from a probability density uniform from 0 to 1). What we will do is write a Monte Carlo simulation (see the Appendix) of the physical process of tossing our two dice six times. In other words, the simulation code will actually "play the game."

The MATLAB code **dtoss.m** (for *double toss*) simulates a game of tossing two fair dice six times and, after each tossing, recording the sum of the dice. At the completion of the sixth toss, the code checks to determine if the first sum is different from each of the next five sums. The code does this for ten million games. (Note: The line numbers at the far left are ***NOT*** part of MATLAB, but have been inserted purely as reference tags.) On a quite ordinary, several years old laptop, the simulation of ten million games (using 120 million random numbers) required less than four seconds.

The code is very elementary and almost (if not totally) self-explanatory. Perhaps the only detail to mention is the double equality (==) in Lines 12 and 16, which is *not* an assignment command (as is the single =), but rather a test for equality of the expressions on each side of the double equal signs.

```
%dtoss.m
01  bingo=0;
02  for loop=1:10000000
03          for toss=1:6
04                  R=6*rand;
05                  d1=floor(R)+1;
06                  R=6*rand;
07                  d2=floor(R)+1;
08                  s(toss)=d1+d2;
09          end
10          match=0;
11          for k=2:6
12                  if s(1)==s(k)
13                  match=1;
14                  end
15          end
16          if match==0
17                  bingo=bingo+1;
18          end
19  end
20  bingo/loop
```

Line 01 initializes the variable *bingo* to zero, where the value of *bingo* is the current number of games in which the first sum is *different* from any of the last five sums. Lines 02 and 19 define the loop that simulates ten million games. Lines 03 and 09 define the loop that plays an individual game. Lines 04 and 05 simulate a toss of the first die, with R being a random number from the interval $0 < R < 6$ (you will recall from the Appendix that the end points, 0 and 6, are *not* possible values for R). Line 05 rounds *down* to the greatest integer *less than* R and then adds 1 (for example, $\text{floor}(4.3) = 4$ and so $\text{floor}(4.3) + 1 = 5$). Thus, $d1$ is a random integer from 1 to 6. Lines 06 and 07 do the same for the second die ($d2$). Line 08 computes the sum of the two dice and stores the result in the six-element vector s. When the six double tosses are finished, the variable *match* is set to zero in Line 10, and then Lines 11 through 15 check to see if $s(1)$ equals any of the last five sums. If a match is found, *match* is set equal to 1. If no match is found, *match* remains equal to 0. Lines 16 through 18 increment *bingo* by 1 if *match* is zero. Line 20 computes, after the ten millionth game is completed, the probability of no match.

2.4 The Computer Solution Shows Its Real Value

Running **dtoss.m** several times produced estimates for the probability of no match that vary from 0.5611287 to 0.561623. That's pretty good agreement with theory (*perfect* agreement, in fact, in the first three decimal digits) and so you report this to your boss with confidence. After reading your report, she looks up with a smile and says, "Nice work, great job, but I need just a bit more. I would like to offer this game to our guests in a slightly different form. Instead of keeping track of *sums*, what if we use the *positive difference* of the two dice? That way, there are only six possible results on each toss, instead of eleven, and I think that will make the game less complicated for players."

This, of course, is an easy request to fill because all you need do is make one simple change in **dtoss.m**. In Line 08 change $s(toss) = d1 + d2$ to $s(toss) = abs(d1 - d2)$, where *abs* is the MATLAB command for computing the *absolute value* of $d1 - d2$ (to get the *positive* difference). When you mention this to your boss, she is not quite as sure about computers as you are. "Sure," she says, "go ahead and run your program and see what it says, but I would still like to see a purely theoretical analysis, too. We are going to use *your* answer to set the odds[3] for this game, and there could be a lot of money riding on *you* being right. Of course, having the theory and the computer agree *would* be the best outcome of all."

TABLE 2.2.2. All Possible Results of Double Tossing Two Dice and Recording Their Positive Difference

Difference	# of ways	How the difference can occur
0	6	(1,1), (2,2), (3,3), (4,4), (5,5), (6,6)
1	10	(1,2), (2,1), (2,3), (3,2), (3,4), (4,3), (5,4), (4,5), (5,6), (6,5)
2	8	(1,3), (3,1), (2,4), (4,2), (3,5), (5,3), (4,6), (6,4)
3	6	(1,4), (4,1), (2,5), (5,2), (3,6), (6,3)
4	4	(1,5), (5,1), (2,6), (6,2)
5	2	(1,6), (6,1)

Not missing the emphasis on *you*, you agree with your boss and rush back to your office to get to work. Running **dtoss.m** several times, with the change mentioned earlier to Line 08, the code returns estimates with the first three decimal digits always at 0.355 for the probability of no match, considerably different from the case of computing sums. So, maybe your boss is right; perhaps a theoretical analysis wouldn't be a waste of time, just to be sure the code is correct. So, to start as before, you construct Table 2.2.2, which shows all the ways the various positive differences can occur.

From the table you immediately see that the total probability the first difference does not match any of the next five differences is given by

$$\left(\frac{6}{36}\right)\left(\frac{30}{36}\right)^5+\left(\frac{10}{36}\right)\left(\frac{26}{36}\right)^5+\left(\frac{8}{36}\right)\left(\frac{28}{36}\right)^5+\left(\frac{6}{36}\right)\left(\frac{30}{36}\right)^5$$
$$+\left(\frac{4}{36}\right)\left(\frac{32}{36}\right)^5+\left(\frac{2}{36}\right)\left(\frac{34}{36}\right)^5$$

$$=\frac{6(30)^5+10(26)^5+8(28)^5+6(30)^5+4(32)^5+2(34)^5}{36^6}$$

$$=\frac{(145{,}800{,}000)+(118{,}813{,}760)+(137{,}682{,}944)+(145{,}800{,}000)+(134{,}217{,}728)+(90{,}870{,}848)}{2{,}176{,}782{,}336}$$

$$=\frac{773{,}185{,}280}{2{,}176{,}782{,}336}$$

or, as the Euclidean algorithm (note 2) tells us, the greatest common divisor for the numerator and the denominator is 256, and the probability we are after is

$$\frac{3,020,255}{8,503,056}=0.3551964\ldots$$

which is in excellent agreement with **dtoss.m**. For this new version of the game, the probability of a player winning (getting a match) is 0.645 and so for every 1,000 games the casino would take in $5,000 and pay out $6,450. That is not going to work, but suppose we redefine a win to be *not* getting a match. Then, for every 1,000 games the casino pays out $3,550 with a take in of $5,000. That is a lot better!

With both the theoretical calculations and the computer simulation result in hand, you return to your boss' office where, after a few words of appreciation, she lays a new twist on you. "Well," she says, "all this is just swell but, after you left, I had another idea. Throwing the dice six times takes too long; I want to move things along a bit faster. So, how about shortening the games to a total of *five* tosses? That is, I want you to determine the probability the first difference is unmatched in the next *four* tosses. That should give the player less opportunity to have a match, right? Well, you work the numbers out, okay? And don't bother with a theoretical analysis—your work has convinced me the computer simulation is giving good answers. Report back as soon as possible."

This is, of course, music to your ears, as now all you need do is change Line 03 in **dtoss.m** from *for toss* = 1:6 to *for toss* = 1:5 and Line 11 from *for k* = 2:6 to *for k* = 2:5. With those changes in place, **dtoss.m** tells you the probability of there being no match of the first difference with any of the differences on the next four tosses is 0.431. Your boss didn't ask for a theoretical analysis, but using Table 2.2.2 should allow you to easily confirm this value. As you head back to your boss' office you wonder what she will come up with next—but whatever it is, you're confident **dtoss.m** can handle it and so this is a good place to move on to the next chapter.

3

The Carpenter's Problem

3.1 A "Stop-You-in-Your-Tracks" Math Problem

It's a rare high school math aficionado who hasn't heard this little gem. A hunter leaves his camp to go bear hunting. He walks 15 miles due south, then 15 miles due east, where he spots a bear and shoots it. He then walks 15 miles due north and thus returns directly to his camp. What was the color of the bear?

For most, this is a *What!?* question when they first encounter it. In fact, it is not a joke but rather a perfectly well-defined problem in spherical geometry (and, perhaps, just a touch of geography). The standard explanation is that such a walk can occur only if the hunter's camp is at the North Pole, and so the bear must have been a polar bear and therefore was white. That seems pretty convincing but, actually, it is not an entirely complete line of correct reasoning. That's because such a walk *is* possible in a region near the South Pole, too, but since there aren't any bears of any color there, the walk must have been in the northern region[1] and so the original answer follows.

In this chapter, we will study the following problem (and a variant of it) that, while perhaps not quite as outrageous as the hunter's problem, might still (initially, at least) give you pause.

A carpenter has two boards, one of length a and another of length b, $a < b$. Suppose she randomly[2] cuts each board into two pieces, and then randomly selects three of the four pieces. What is the probability she can construct a triangular window frame from those three pieces?

3.2 The Triangle Inequality

So, you might well ask where to begin after reading that. A good place to start is with a sketch like Figure 3.2.1. In the top half of the figure we have the two boards cut into lengths x and $a-x$, and into lengths y and $b-y$, where x is uniform over 0 to a, and y is uniform over 0 to b. Notice that to pick three pieces from four pierces means the carpenter must select both pieces from one cut board, and the third piece from one of the two pieces from the other cut board. That means there are *four* possible configurations for a triangular frame, shown in the bottom half of Figure 3.2.1.

The key observation that unlocks this problem is to recall the *triangle inequality*, which says the sum of any two sides of a triangle is greater than the third side. This is, of course, simply a statement of the well-known fact that the shortest distance between two points is a straight line. Taking, in turn, each of the four configurations of Figure 3.2.1, let's apply the triangle inequality and see what we get.

For configuration I, we have $x+y>a-x$, $x+(a-x)>y$, and $(a-x)+y>x$, which reduce to $y>a-2x$, $a>y$, and $y>2x-a$. In Figure 3.2.2 these three inequalities have been drawn on the entire sample space (see the Appendix) for x and y, and we see that the points in the shaded region simultaneously satisfy all three inequalities, while the points not in the shaded region do not. Since x and y are each uniform random quantities, the probability of the shaded region is simply the ratio of the area of the shaded region to the area of the sample space (this is called geometric probability). So,

$$Prob_I = \frac{\frac{1}{2}a^2}{ab} = \frac{a}{2b}.$$

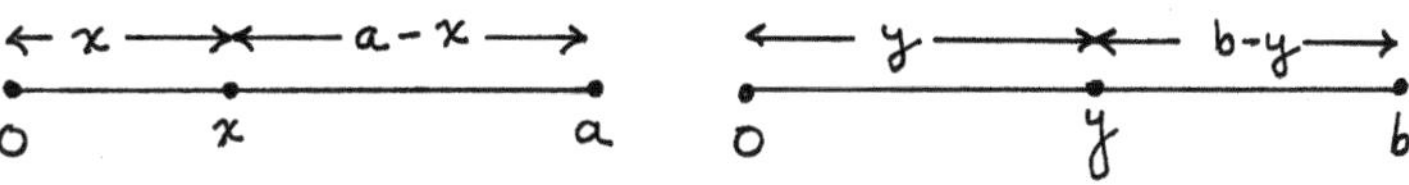

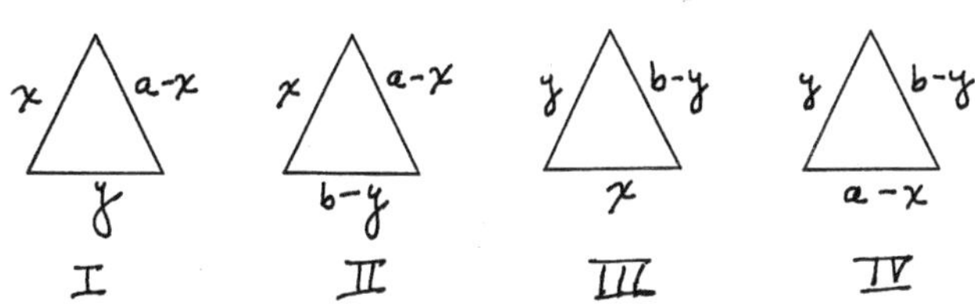

FIGURE 3.2.1. The four possible configurations for a triangular frame.

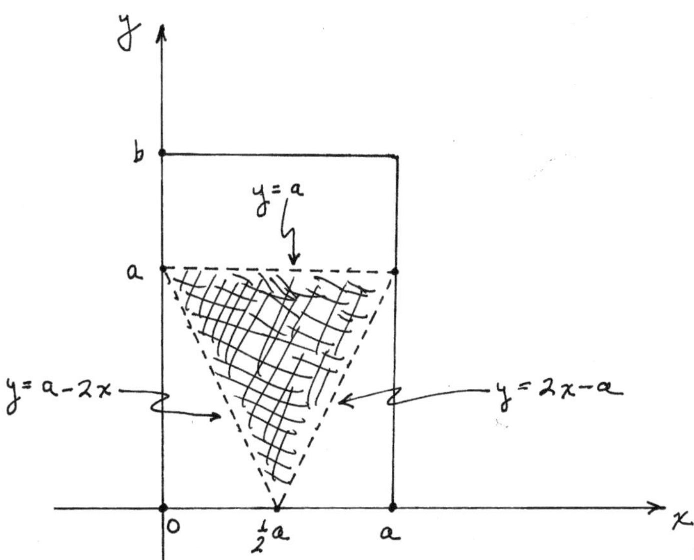

FIGURE 3.2.2. All three inequalities for configuration I hold in the shaded region.

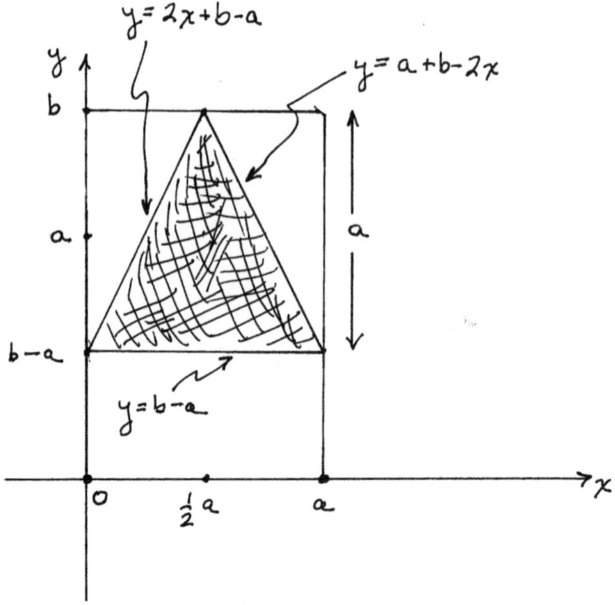

FIGURE 3.2.3. All three inequalities for configuration II hold in the shaded region.

Turning next to configuration II, applying the triangle inequality gives $x+(a-x)>b-y$, $(a-x)+(b-y)>x$, and $(b-y)+x>a-x$. You can do the easy steps that show these reduce to $y>b-a$, $y<a+b-2x$, and $y<2x+b-a$. Figure 3.2.3 shows the region in sample space where all three of these inequalities simultaneously hold. Using geometric probability (ratio of areas) we see that

$$Prob_{II}=\frac{\frac{1}{2}a^2}{ab}=\frac{a}{2b}.$$

Moving on to configuration III, the triangle inequality states $x+y>b-y$, $y+(b-y)>x$, and $(b-y)+x>y$. These statements quickly reduce to $y>\frac{1}{2}(b-x)$, $x<b$ (which is, of course, always

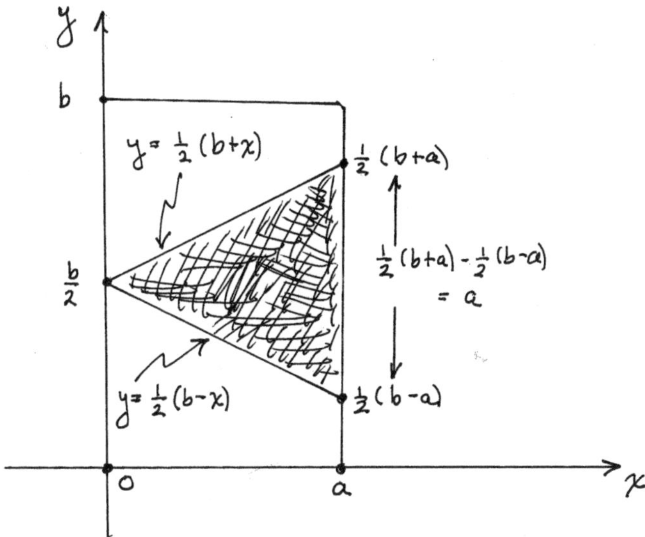

FIGURE 3.2.4. All three inequalities for configuration III hold in the shaded region.

the case as $x<a<b$), and $y<\frac{1}{2}(b+x)$. Figure 3.2.4 shows the region in sample space where all three inequalities simultaneously hold. Thus, by geometric probability,

$$Prob_{III}=\frac{\frac{1}{2}a^2}{ab}=\frac{a}{2b}.$$

Finally, for configuration IV, the triangle inequality states $(a-x)+y>b-y$, $y+(b-y)>a-x$, and $(b-y)+(a-x)>y$, which quickly reduce to $y>\frac{x+(b-a)}{2}$, $b-a>-x$ (which is always true as $b-a>0$ and $-x<0$), and $y<\frac{(b+a)-x}{2}$. Figure 3.2.5 shows the region in sample space where all three inequalities simultaneously hold. Thus, by geometric probability,

$$Prob_{IV}=\frac{\frac{1}{2}a^2}{ab}=\frac{a}{2b}.$$

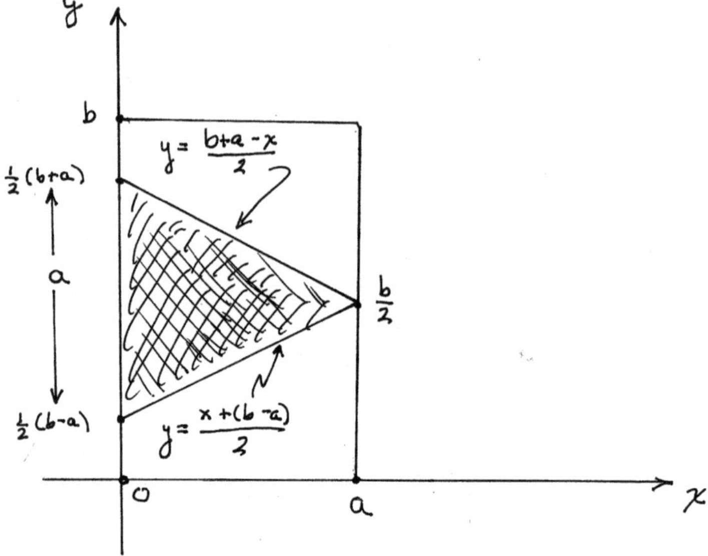

FIGURE 3.2.5. All three inequalities for configuration IV hold in the shaded region.

You have surely noticed by now that each of the four configurations results in the *same probability,* and thus it doesn't matter which one actually occurs. So, for example, if $a=7$ feet and $b=9$ feet, then the probability of a triangular frame is $\frac{7}{18}\approx 0.3889$. A more detailed way to think of this is to observe that nothing makes one of the four possible configurations any more or less likely to occur than any other. That is, each configuration has probability $\frac{1}{4}$ of occurring. So, suppose we perform the details of the carpenter's task a very large number of N times. We would then expect that each configuration would occur about $\frac{1}{4}N$ times. Since the probability of a triangular frame is $\frac{a}{2b}$ from each configuration, we expect a total of

$$\frac{1}{4}N\frac{a}{2b}+\frac{1}{4}N\frac{a}{2b}+\frac{1}{4}N\frac{a}{2b}+\frac{1}{4}N\frac{a}{2b}=N\frac{a}{2b}$$

triangular frames to occur in the N total tries. Thus, the overall probability of a triangular frame is

$$\frac{N\frac{a}{2b}}{N}=\frac{a}{2b}.$$

3.3 Monte Carlo Computer Solution

So, by using a little bit of elementary probability theory we have solved the carpenter's problem. But suppose your knowledge of probability theory is limited to simply knowing that your computer has a uniform (0 to 1) random number generator command (*rand* in MATLAB). Well then, you are *still* good to go and able to solve the carpenter's problem, for any given values of a and b, using a computer code like **carpenter.m** (which is simply a brute-force application of the triangle inequality). Running that code numerous times (for $a = 7$ and $b = 9$, for example) within seconds returned a probability of 0.3889, in excellent agreement with theory. Here is a walk-through of how that code works.

```
%carpenter.m
a=7;b=9;bingo=0;
for loop=1:100000000
     x=a*rand;y=b*rand;
     (1)=x;l(2)=a-l(1);l(3)=y;l(4)=b-l(3);
     p=rand;
     if p<0.25
          side1=l(2);side2=l(3);side3=l(4);
     elseif p<0.5
          side1=l(1);side2=l(3);side3=l(4);
     elseif p<0.75
          side1=l(1);side2=l(2);side3=l(4);
     else
          side1=l(1);side2=l(2);side3-l(3);
     end
```

```
15        if side1+side2>side3
16             if side2+side3>side1
17                  if side3+side1>side2
18                       bingo=bingo+1;
19                  end
20              end
21         end
22    end
23    bingo/loop
```

Line 01 sets the values of *a* and *b*, and initializes the value of *bingo* to zero (*bingo* will, at the end of the simulation, be the number of triangular frames created in 100 million tries). The *for/end* loop defined by Lines 02 and 22 controls the 100 million tries. With the top half of Figure 3.2.1 in mind, Line 03 determines the values of *x* and *y*, the locations of the cuts in the *a* board and the *b* board, respectively. Line 04 sets the four elements of the *l*-vector to the lengths of the resulting four pieces after completing a double cut. Lines 05 through 14 decide which configuration in the lower half of Figure 3.2.1 to simulate, with each configuration having probability $\frac{1}{4}$ of being selected (Lines 06 and 07 select configuration IV, Lines 08 and 09 select configuration III, Lines 10 and 11 select configuration II, and Lines 12 and 13 select configuration I). Lines 15 through 21 apply the triangle inequality to the selected configuration, with Line 18 incrementing *bingo* by 1 each time all three inequalities are satisfied. Line 23 gives the probability of having a triangular frame after exiting the *for/end* loop of Lines 02 and 22.

3.4 A Somewhat More Complicated Version of the Problem

All of what we have just done probably strikes you (since you have seen how the reasoning goes) as pretty straightforward. Now consider a (perhaps) seemingly minor variant of the carpenter's problem. We will still be able to solve it theoretically, but only at the cost of having to upgrade the firepower of the probability theory needed. To solve this new problem with a computer code, however, will require only simple (and obvious) alterations to **carpenter.m**.

The variant opens with our carpenter now having just a single board, which we will take as having length l. She then cuts the board twice at random (I will soon make this more precise), resulting in three pieces. What is the probability those three pieces form a triangular frame? There are, at least, two obvious ways to make the two cuts, which I will call Method 1 and Method 2.

Method 1: Mark two points x and y on the board, *independently*, as measured from the left end of the board. Cut the board at those two points. That is, the two marked points are each uniform over the interval 0 to l.

Method 2: Cut the board at random, giving two pieces. Then, randomly select one of those two pieces and randomly cut it into two pieces.

Now, just for fun, before doing any calculations, what does your intuition tell you about these two methods? Do you think they are equivalent in the sense of giving the same probability? Or, if not equivalent, which method has the greater probability of resulting in a triangular frame? Got it? Okay, let's see how good your intuition is.

The theoretical analysis of Method 1 is pretty much like our analysis for the original carpenter's problem, with an additional twist: an appreciation of the subtle nature of the sample space for our new problem. In addition to the three triangle inequalities, we now have to add a fourth one that addresses the two possibilities of $x < y$ and $x > y$. Since x and y are each chosen independently, both possibilities need to be considered (which I will call case a and case b, respectively), as shown in Figure 3.4.1. Nothing favors one case over the other, so it seems reasonable to take the probability of each to be $\frac{1}{2}$.

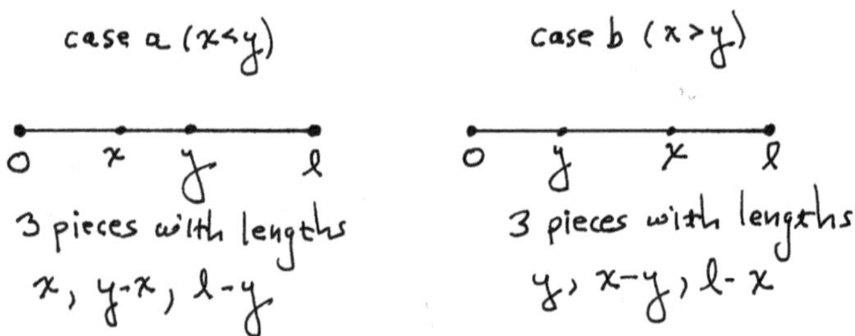

FIGURE 3.4.1. The two versions of Method 1.

For case a, the *four* inequalities we have to work with are $x+(y-x)>l-y$, $(y-x)+(l-y)>x$, $(l-y)+x>y-x$, and $x<y$. These quickly reduce to $y>\frac{l}{2}$, $x<\frac{l}{2}$, $y<x+\frac{l}{2}$, and $x<y$. Plotting the first three inequalities on an l -by-l square, we arrive at Figure 3.4.2, which looks much like the figures from the original carpenter's problem. However, the sample space for our new problem is not l -by-l, but rather is only the top diagonal half of the l -by-l square because of the last inequality $x<y$. That is, the area of sample space in case a is not l^2 but rather is $\frac{1}{2}l^2$. Since the area of the shaded region of Figure 3.4.2 (where the triangle inequalities hold) is $\frac{l^2}{8}$, then by geometric probability we have the probability of a triangular frame for case a as $\frac{\frac{l^2}{8}}{\frac{1}{2}l^2}=\frac{1}{4}$.

For case b of Method 1, the four inequalities we have are $y+(x-y)>l-x$, $(x-y)+(l-x)>y$, $(l-x)+y>x-y$, and $x>y$. These quickly reduce to $x>\frac{l}{2}$, $y<\frac{l}{2}$, $y>x-\frac{l}{2}$, and $x>y$, which produce Figure 3.4.3, with the shaded region showing where the three triangle inequalities hold. The fourth inequality defines the sample space as having area $\frac{1}{2}l^2$. So, the probability of a triangular frame for case b is, as in case a, $\frac{\frac{l^2}{8}}{\frac{1}{2}l^2}=\frac{1}{4}$.

Using the same sort of argument as the one at the end of section 3.2, the overall probability of getting a triangular frame using Method 1 is $\frac{1}{4}$. The computer code **method1.m** (the operation of which should now be transparently obvious) confirms this:

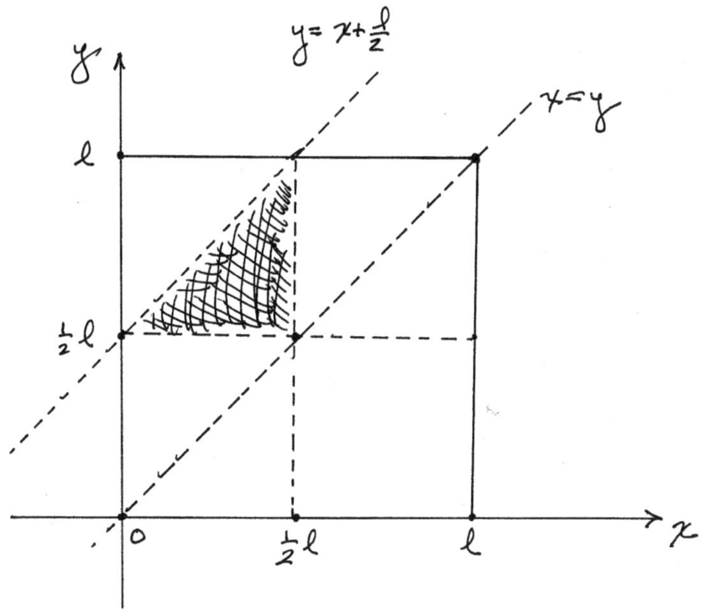

FIGURE 3.4.2. Case a for Method 1 ($x < y$).

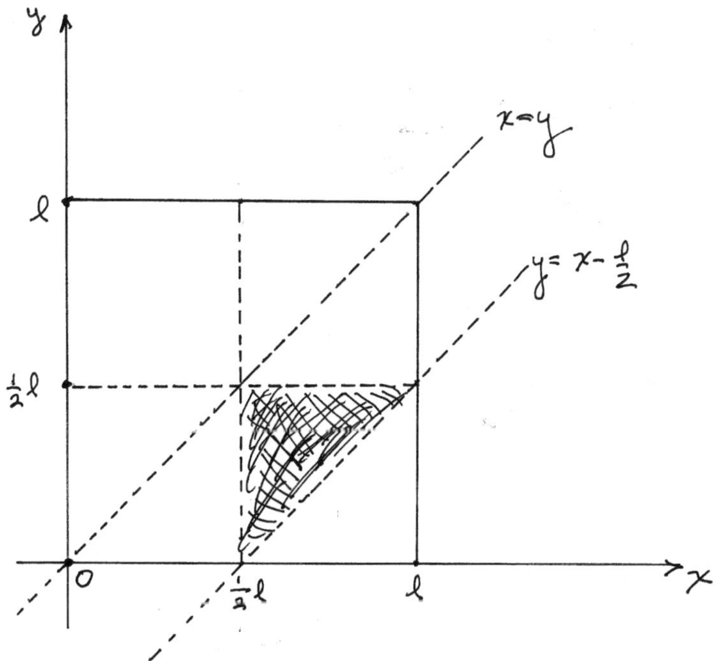

FIGURE 3.4.3. Case b for Method 1 ($x > y$).

simulation of 100 million double cuts (using a total of 200 million random numbers!) gave a probability of 0.24994.[3]

```
%method1.m
bingo=0;
for loop=1:100000000
     x=rand;y=rand;
     if x<y
          side1=x;side2=y-x;side3 = 1-y;
     else
          side1=y;side2=x-y;side3 = 1-x;
     end
     if side1+side2>side3
          if side2+side3>side1
               if side3+side1>side2
                    bingo=bingo+1;
               end
          end
     end
end
bingo/loop
```

We next turn our attention to Method 2, which also has a case a and case b (see Figure 3.4.4). With Method 2, however, an entirely new complication occurs. The top half of the figure shows an initially intact board of length 1 (see note 3 again) cut into two pieces, of lengths x and $1-x$, where x (the location of the first cut) is uniform over 0 to 1. The bottom half of the figure then shows the two possibilities for the second cut. In case a, the piece of length x is cut at y (with y uniform from 0 to x), and in case b the piece of length $1-x$ is cut at y (with y uniform from x to 1). Notice, *carefully*, that x and y are *not* independent, because in each case knowledge of x gives us information about y. In case a, once we know x it then immediately follows that $y<x$, while in case b we would know $y>x$. This will soon have big implications to consider.

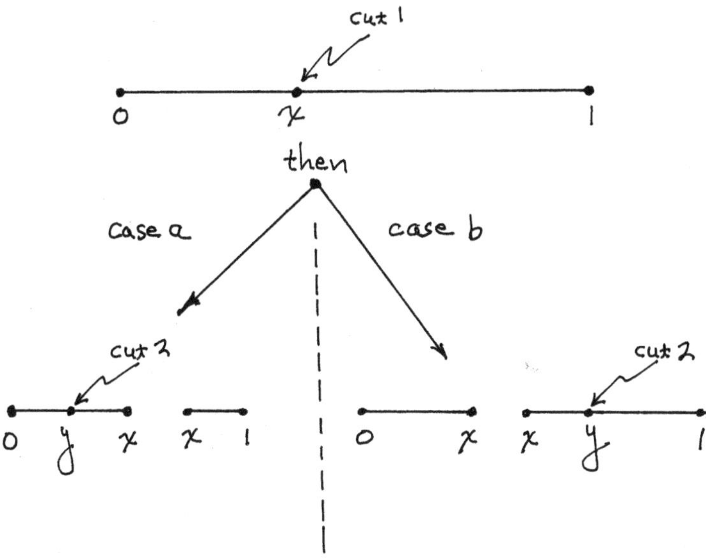

FIGURE 3.4.4. The two equally probable possibilities for Method 2.

An analysis of case a begins just as we have done before. The triangle inequalities are $y+(x-y)>1-x$, $(x-y)+(1-x)>y$, and $(1-x)+y>x-y$, which reduce to $x>\frac{1}{2}$, $y<\frac{1}{2}$, and $y>x-\frac{1}{2}$. When plotted on sample space we get just what we got for case b of Method 1 (see Figure 3.4.3) with l set to 1. The big difference in that earlier analysis compared to this one is that in Method 1 x and y *are* independent, so we could use geometric probability to calculate the probability of the shaded region as the ratio of two areas. But now, with Method 2, x and y are *not* independent (as argued in the previous paragraph), and so geometric probability doesn't apply. What a sad state of affairs, or so it seems.

Well, there *is* a way to directly calculate the probability of the shaded region even when x and y are dependent but, before we see how that works, let's go right to a computer simulation that gets us a numerical answer even if our probability knowledge is lacking (a major theme of this book). The code **method2.m** does the trick, and a walk-through of it shows how the code works. Most of it, actually, should look pretty familiar to you by now, particularly the triangle inequalities in Lines 16, 17, and 18.

```
%method2.m
01  bingo=0;
02  for loop=1:100000000
03        x=rand;
04        decision=rand;
05        if decision<0.5
06              y=x*rand;
07              side1=y;
08              side2=x-y;
09              side3 = 1-x;
10        else
11              y=x+(1-x)*rand;
12              side1=x;
13              side2=y-x;
14              side3 = 1-y;
15        end
16        if side1+side2>side3
17              if side2+side3>side1
18                    if side3+side1>side2
19                          bingo=bingo+1;
20                    end
21              end
22        end
23  end
24  bingo/loop
```

The code simulates Method 2 one hundred million times (via the loop defined by Lines 02 and 23), with Line 03 making the first cut (assigning a random value to x). A second random number in Line 04 determines the case that will be treated (case a in Lines 07 to 09, or case b in Lines 12 to 14). Lines 06 and 11 are the ones that make this code something new, as they determine the location of the second cut (the value of y) for the two cases. Those two lines handle the new issue we face about how to theoretically handle the fact that x and y are not independent.

Running **method2.m** numerous times gives the consistent result that the probability of a triangular frame is 0.1931, which is significantly less than the probability for Method 1. The two methods are clearly *not* equivalent. (Is that what your intuition told you?) However, we are still faced with the issue of confirming, theoretically, that the seemingly unremarkable number 0.1931 is correct. So, here is how to *directly calculate* the exact probability of the region in sample space where all the triangle inequalities are satisfied.

In Figure 3.4.5 I have redrawn Figure 3.4.3 (with $l = 1$) to show the region for case b of Method 1 divided into vertical strips of differential width dx. To find the probability of the region, we integrate the joint probability density function (pdf) of x and y over that region (see the Appendix). Writing the joint pdf as $f(x, y)$, we start with the fundamental relation (again, see the Appendix)

$$f_{Y|X}(y\,|\,x) = \frac{f(x, y)}{f_X(x)}$$

where $f_X(x)$ and $f_{Y|X}(y|x)$ denote, respectively, the pdf of the random variable X, and the *conditional* pdf of the random variable Y *given* the value of X.

Since the first cut is uniformly random from 0 to 1, we have $f_X(x) = 1$, and since in case a the second cut is uniformly random from 0 to x, we have $f_{Y|X}(y\,|\,x) = \frac{1}{x}$ Thus, the joint pdf is

$$f(x, y) = f_{Y|X}(y\,|\,x)\,f_X(x) = \frac{1}{x}.$$

The probability of the region of interest (the one with vertical stripes) is given by the double integral

$$Prob_a = \iint_{region\ of\ interest} f(x, y)\,dy\,dx.$$

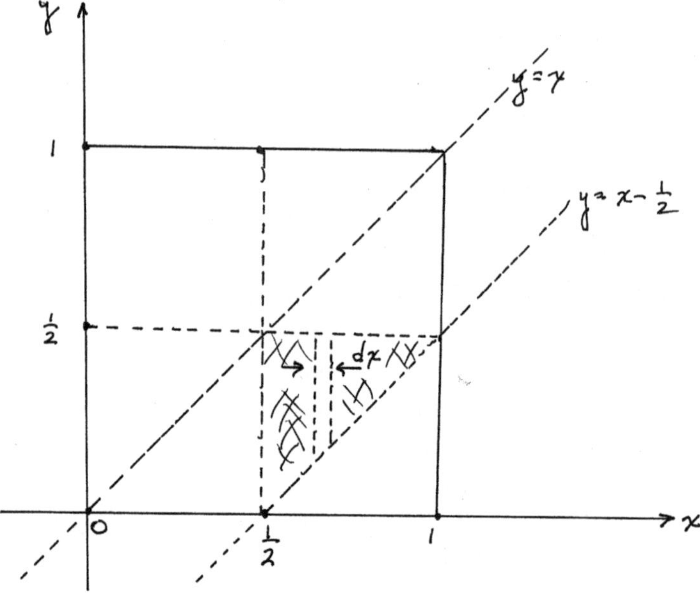

FIGURE 3.4.5. Integrating the joint pdf of x and y in case a of Method 2.

We can put specific limits on the x and y integrals by looking at Figure 3.4.5 and observing how, for a given x (which varies from $\frac{1}{2}$ to 1), y varies along a vertical strip from $x-\frac{1}{2}$ to $\frac{1}{2}$, and so

$$Prob_a = \int_{1/2}^{1}\left[\int_{x-1/2}^{1/2}\frac{1}{x}dy\right]dx = \int_{1/2}^{1}\left[\int_{x-1/2}^{1/2}dy\right]dx$$
$$= \int_{1/2}^{1}\frac{1}{x}\left\{\frac{1}{2}-(x-1/2)dx\right\} = \int_{1/2}^{1}\frac{1}{x}(1-x)dx$$
$$= \int_{1/2}^{1}\frac{dx}{x} - \int_{\frac{1}{2}}^{1}dx = ln(x)\big|_{\frac{1}{2}}^{1} - \left(1-\frac{1}{2}\right)$$
$$= ln(1) - ln\left(\frac{1}{2}\right) - 1/2 = -ln\left(\frac{1}{2}\right) - \frac{1}{2}$$
$$= -ln(1) + ln(2) - \frac{1}{2} = ln(2) - \frac{1}{2} = 0.1931\ldots.$$

This is in excellent agreement with **method2.m**'s result.

We aren't quite done yet, of course, as we also have to work out case b. You can fill in the details, but if you write out the triangle inequalities for case b of Method 2 you will see they define the same region (for $l = 1$) that we got for case a of Method 1. Since in case b of Method 2 we have $f_X(x) = 1$ and $f_{Y|X}(y|x) = \frac{1}{1-x}$, then the joint pdf to be integrated over the region is $f(x, y) = \frac{1}{1-x}$. This leads to the double integral

$$Prob_b = \int_0^{1/2} \left[\int_{1/2}^{x+1/2} \frac{1}{1-x} dy \right] dx$$

which, if you do the integrations correctly, again leads to $ln(2) - \frac{1}{2} = 0.1931\ldots$. Give it a try!

Remember, the central point of all this is *not* that theory is unnecessary, but rather that if your math isn't quite up to the demands made by theory, you may still be able to study your problem and get results on a computer. Best of all, of course, would be to have *both* tools in your bag of techniques.

4

Wi-Fi Coverage and Antisubmarine Warfare Are the Same Math Problem

4.1 Introduction

If you do a Google search on the Internet of the phrase *area of intersecting circles* (or some variation of it), a lot (*millions*!) of hits will result. It seems to be referencing a very popular math question! It certainly is a challenging one. To find the intersection (or common) area of two circles in the same plane, given the locations of their centers and their radii, is not a trivial exercise. For three circles the math becomes exponentially more involved, for four circles the calculations become *extremely* difficult, and for five or more circles only madness happens.

That's the case, anyway, if one tries a totally analytic solution involving multiple, simultaneous circle equations. However, many of that sort of problem have sufficient structure to be solved by recognizing some sort of symmetry that allows a one-off trick to work. Such problems are understandably popular with high school math teachers because they make great challenge questions for even the best students. As an example, Figure 4.1.1 shows three circles with identical radii R, positioned such that the center of each (labeled A, B, and C) is on the intersection of the circumfer-

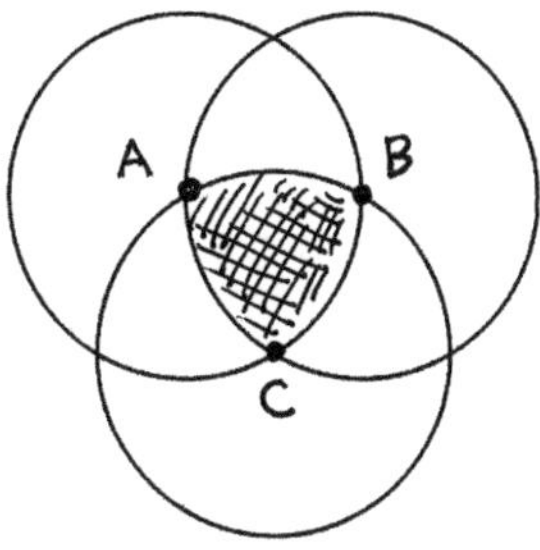

FIGURE 4.1.1. The radius of each circle is R.

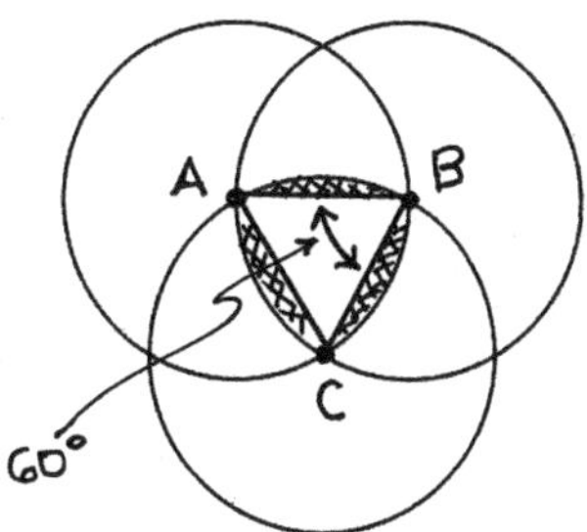

FIGURE 4.1.2. The common area is the sum of the areas of the triangle and three spherical caps (shown shaded).

ences of the other two circles. The problem is to calculate the common (shaded) area of the three circles.

The trick that solves this problem is shown in Figure 4.1.2, which makes it obvious that the common area is the sum of the area of an equilateral triangle (the triangle ABC is equilateral because its three sides are each equal to R, the radius of each circle) and the areas of three circular "caps" (shown as shaded). These areas are easily calculated using high school geometry. The triangle area is particularly easy to calculate, and I will leave it to you to show that the area is $\frac{\sqrt{3}}{4}R^2$ (just remember that the area of a triangle is one half the base times the height, and use the Pythagorean theorem).

The observation that completes the trick is that since the interior angles of an equilateral triangle are each 60°, then the area of the triangle and one circular cap is one-sixth of the area of a circle (look again at Figure 4.1.2). That is, if the area of a circular cap is written as A_{cap}, then

$$\frac{\sqrt{3}}{4}R^2 + A_{cap} = \frac{1}{6}\pi R^2$$

or,

$$A_{cap} = \frac{1}{6}\pi R^2 - \frac{\sqrt{3}}{4}R^2 = R^2\left(\frac{\pi}{6} - \frac{\sqrt{3}}{4}\right).$$

So, the answer to our problem is, as stated before, the sum of the triangle area and three cap areas, that is,

$$\frac{\sqrt{3}}{4}R^2 + 3R^2\left(\frac{\pi}{6} - \frac{\sqrt{3}}{4}\right) = \frac{1}{2}(\pi - \sqrt{3})R^2.$$

This is, without question, a fun problem, but one that has been clearly put together to have a nice (if tricky) solution. Even a slight change in the details of the problem will, alas, render our trick useless. There may, in fact, not be *any* trick that will then work. What do we do in that case?

In the rest of the first half of this chapter we will address the following two issues: (1) *Why* would anyone care about such a calculation? (In the discussions of the problems I have found on the Internet, there is typically—as with the preceding example—no physical motivation provided.) (2) With a computer, a very simple algorithm (that will be developed) allows one to get a result for *any* number of circles of arbitrary location and radii.

In second half of the chapter, we will cover a *real-world* extension of the circle overlap problem into the third dimension: that is, the problem of intersecting *volumes of spheres* of arbitrary location and radii. The algorithm developed for the circle overlap case extends, in a quite simple and obvious way, to the sphere overlap one.

4.2 The Wi-Fi Problem

A Wi-Fi (for *wireless fidelity*) system is based on the presence of a radio-frequency signal that exists in a circular area of some finite radius, centered on an electronic gadget that generates the signal. Devices like cellphones and laptop computers can connect to the gadget via the signal (and from there to the Internet or commercial telephone networks). Everybody reading this book is virtually certain to have used Wi-Fi in a coffee shop, a library, or other public building, as well as in their own home.

As a specific example, consider the problem faced by the owner of a circular restaurant who wants to provide Wi-Fi service to his customers. The radius of his restaurant is significantly larger than the radius of the circular area occupied by the signal generated by any one of his available Wi-Fi gadgets, and so he decides to simply install multiple (four) identical gadgets, as shown in Figure 4.2.1. That figure, to be really specific, supposes the radius of the restaurant is two times the signal radius of a gadget (in the figure, the clear circles are where the signal of a gadget is present, while in the shaded areas no Wi-Fi signal is present). The owner would like to know the fraction of his total floor space that will have no Wi-Fi.

The answer to the owner's question can be calculated *exactly* using no more than high school geometry, and by the end of this section you will see how to do that (you can try to do it yourself right now). But first, here is a very simple computer solution (the correctness of which we will verify with the theoretical solution). To begin, notice that the answer (a *fraction*) will be independent of the actual values of the restaurant floor area or the value of the Wi-Fi signal radius of a gadget. All that will matter is the 2-to-1 ratio of the restaurant radius to the signal radius. So, let's pick a particularly convenient value for the Wi-Fi radius (say, $r=1$) and thus the radius of the circular restaurant is $R=2$. If we put the center of the restaurant at the origin of a rectangular coordinate system, then from Figure 4.2.1 you can see that the Wi-Fi signal generators are located at $(1,0)$, $(0,1)$, $(-1,0)$, and $(0,-1)$.

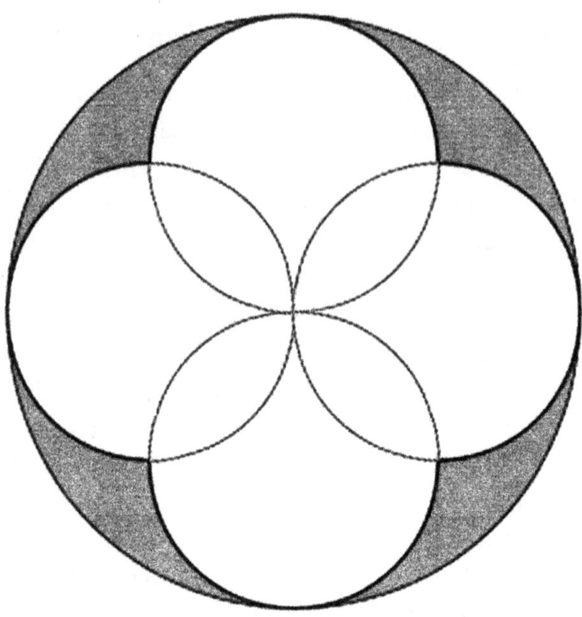

FIGURE 4.2.1. Wi-Fi in a circular restaurant with four signal generators.

Now, imagine that we place the restaurant circle inside a square of edge length 4, with the square also centered on the coordinate axes. Thus, the edges of the square extend from −2 to 2, in both the horizontal and vertical directions. We next generate a very large number of points, at random, over the interior of the square. If (x, y) is the location of any such point, then the point is not only inside the square but also inside the restaurant circle *if* $x^2 + y^2 < (2)^2 = 4$. Further, if the distance between such a point and *any* of the signal generators is less than 1, then that point is a location where Wi-Fi service will be available. Figure 4.2.2 shows the logic of an algorithm that determines if a randomly selected point from the interior of the square[1] is in the shaded portion of Figure 4.2.1.

Next, imagine the logic of Figure 4.2.2 is placed inside a loop that repeats it a very large number (say, ten million) times. Every

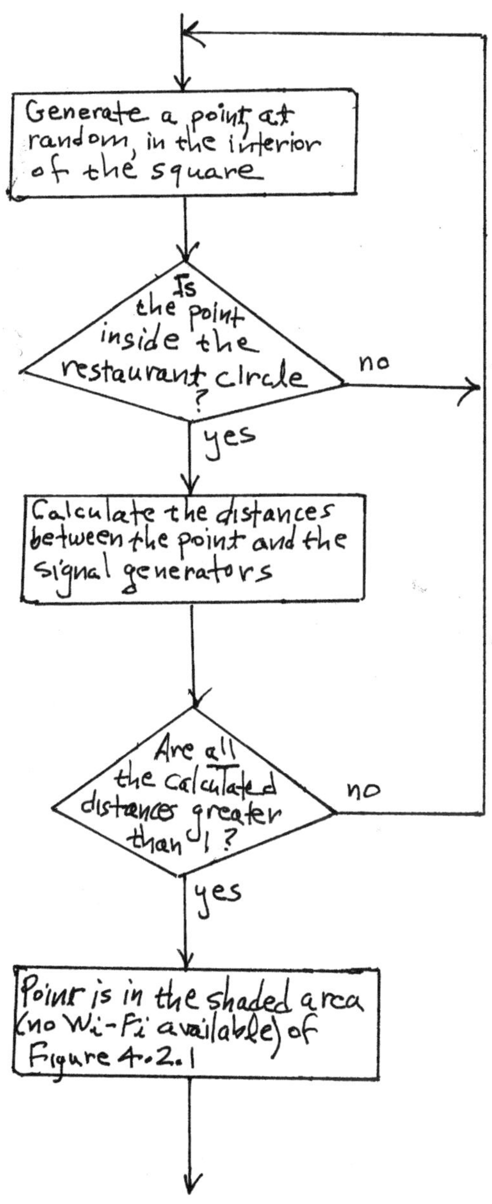

FIGURE 4.2.2. Logic for determining if a random point is in the shaded area of Figure 4.2.1.

time the logic finds a point in the shaded area of Figure 4.2.1, imagine that a variable called *bingo* is incremented by one. If we initialize *bingo* at zero before entering the loop, then when the last pass through the loop is done, *bingo* will equal the total number of points that were in the shaded (no Wi-Fi available) area. Since the fraction *bingo* / 10000000 is a measure of the fractional area of the restaurant circle area where there is no Wi-Fi, we have the answer to the restaurant owner's question. A MATLAB code that does all this is **wifi.m**, and here is how it works.

```
%wifi.m
01  bingo=0;
02  for loop=1:100000000
03        stop=0;
04        while stop==0
05        x=-2+4*rand;y=-2+4*rand;
06          if x^2+y^2<4
07            stop=1;
08          end
09      end
10      d1=(x-1)^2+y^2;
11      d2=x^2+(y-1)^2;
12      d3=(x+1)^2+y^2;
13      d4=x^2+(y+1)^2;
14      if d1>1&d2>1&d3>1&d4>1
15          bingo=bingo+1;
16       end
17  end
18  bingo/loop
```

Line 01 initializes *bingo* to zero, before the loop defined by Lines 02 and 17 is entered. Lines 03 through 09 generate a point at random inside the square $-2<x<2$, $-2<y<2$ that is also in the restaurant circle. Lines 10 through 13 calculate the distances between the point and each of the signal generators. Line 14 determines if the point is within the range of at least one of the signal generators,

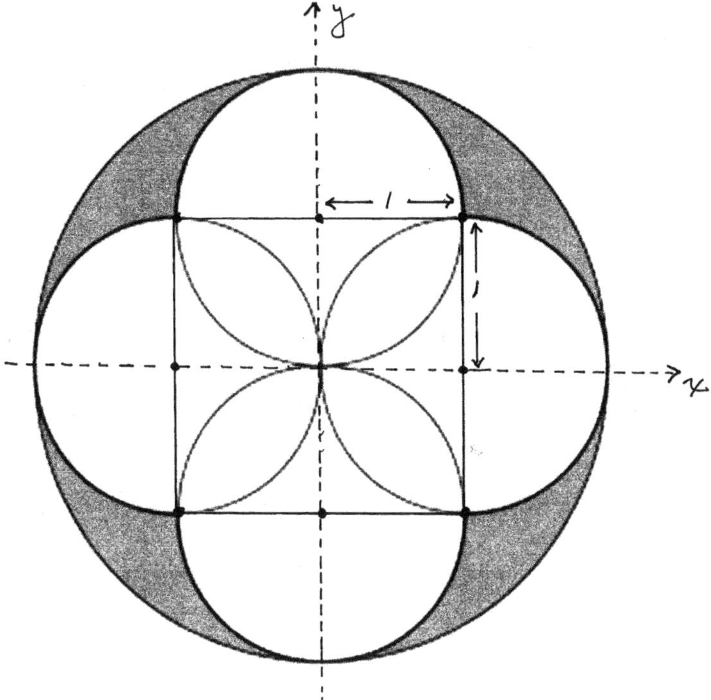

FIGURE 4.2.3. Calculating the exact "no Wi-Fi" area.

using MATLAB's logical-AND operator (&). That is, for Line 15 to be executed, the distances between the point and the signal generators must *all* be greater than the radius of a signal generator (that is, the point is in the no Wi-Fi area of Figure 4.2.1).

Executing **wifi.m** produces, in a matter of mere seconds, the answer we are after: the fraction of the restaurant area without Wi-fi is 0.1817. Well, you say, that's what the *code* says, but is the code right? To answer that, here is how to calculate the *exact* value. Figure 4.2.3 shows Figure 4.2.1 with a central square area of edge length 2 indicated. This designates how the unshaded area (Wi-Fi available) is the sum of the area of the interior square and the areas of four half circles, each with radius 1. That is, the total unshaded area is $(2)^2+4\left[\frac{1}{2}\pi(1)^2\right]=4+2\pi$. Since the area of the restaurant is $\pi(2)^2=4\pi$, we see that the shaded area is $4\pi-(4+2\pi)=2\pi-4$. Thus, fraction of the restaurant with no Wi-Fi is

$$\frac{2\pi-4}{4\pi}=\frac{1}{2}-\frac{1}{\pi}=0.18169.$$

The computer code has done well.

Notice that the code can do lots more than just verify the theoretical calculation. For example, suppose the signal generators are not identical, with one of them having a signal radius only 90 percent of that for the other three generators. Then what is the fractional portion of the restaurant area with no Wi-Fi? Clearly, the fraction will increase, but by how much? The trick of seeing the internal square and half circles doesn't work anymore (be sure you understand this), but in the code all we need do is change Line 14. Because of the circular symmetry of the problem, the answer to our new question doesn't depend on which of the four signal generators is the odd one: let's say it is the generator at $(1,0)$. Then, Line 14 becomes (because $0.9^2=0.81$)

$$\textit{if } d1>0.81\,\&\,d2>1\,\&\,d3>1\,\&\,d4>1.$$

Running **wifi.m** with this change states the fractional area with no Wi-Fi is 0.2055. More generally, all four of the signal generator radii could be different, and in addition the generators could be located anywhere you wish by modifying Lines 10 through 13. It is all grist for the mill for the code.

As a final comment on intersecting circles, a quite interesting example of such a thing occurs in the 1937 novel *Mrs. Miniver* (Harcourt 1940) by Jan Struther (made into a 1942 movie). At one point, Mrs. Miniver is described as follows:

> She saw every relationship as a pair of intersecting circles. It would seem at first glance that the more they overlapped the better the relationship; but this is not so. Beyond a certain point, the law of diminishing returns sets in. . . . Probably perfection is

> reached when the area of [the overlapped region equals that of the non-overlapped region]. On paper there must be some neat mathematical formula for arriving at this; in life, none."[2]

4.3 Intersecting Spheres

After pondering intersecting circles, it seems to be almost impossible to resist wondering about intersecting spheres. That is actually not just an abstract theoretical question, as there *are* physical scenarios where the question of intersecting spheres arises in a natural way. My favorite is one discussed in a beautiful paper[3] in applied mathematics by a University of Wyoming math professor who, under contract with a military aerospace manufacturer (The Martin Company), studied a life-and-death problem in antisubmarine warfare.[4]

The paper's opening sentence is pretty much what you might expect in a math paper—highly technical language describing a situation so general that it is not immediately obvious what might be coming next:

> If a sphere S_1 of radius R is aimed at the center of a sphere S_2 of radius D, several interesting useful probability problems arise.

Professor Guenther, however, quickly sweeps away the abstract and gives his readers a quite specific problem to which his opening words apply.

Imagine, he says, that a hostile submarine has been detected far below the surface of the ocean. The detection can be imagined as having been performed by some advanced technology (for example, a vast undersea network of sound, pressure, and magnetic sensors on the ocean floor that respond to the presence and motion through water of large objects) that we will just accept as being operational. This discovery is transmitted to some weapons platform—an attack submarine or an airborne aircraft—that carries torpedoes with high-explosive warheads. This communication prompts the platform to launch a torpedo attack on the hostile submarine's coordinates where it was first detected. It will, of course, take some interval of time for the torpedo to reach the target's detected position and, during that travel time, either as part of its normal maneuvers or because it knows it is under attack, the hostile submarine has the capability to move some maximum distance D in any direction (which could be many hundreds of yards). In other words, when the torpedo arrives at its aim point (the location of the hostile submarine *when first detected*[5]), the hostile submarine could be anywhere inside a sphere of radius D, centered on its position when first detected. When the torpedo reaches what it "thinks" is its aim point, the warhead is automatically detonated and creates a spherical region of lethal radius R in which nothing can survive.

So, there is our problem: what is the probability that a point (the hostile submarine) that is at a random location inside a sphere of radius D, centered on the initial detection location, is also within distance R of the warhead detonation? That probability is what military analysts call the *kill probability*. The answer depends, obviously, on where the detonation occurs—by intent it is at the initial detection point, but weapons analysts know that will almost certainly not be the case because of what they call *aiming errors*.

Aiming errors are not intentional, but are the unavoidable, cumulative effects of numerous distinct influences, the nature of which may not even be known much less their magnitudes. Such effects for our torpedo could include, for example, variations in the thrust of the torpedoes' propulsion system and the encounter of random crosscurrents during the travel time to the target. Weapons analysts usually assume that the aiming errors, in each of the three spatial axes, are described by independent, zero-mean, normally distributed random variables with equal standard deviations (see the Appendix). Professor Guenther takes the standard deviation of the torpedoes' aiming errors to be the unit distance in his analysis. That is, the values of the radii D and R are in units of the standard deviation of the aiming errors.[6] So, for example, if the aiming error standard deviation is (just to make up a number) 10 feet, and (just to make up a couple more numbers) if D and R are (respectively) 600 feet and 150 feet, then $D=60$ and $R=15$.

With all this in mind, Professor Guenther then performed a very elegant, highly theoretical analysis that requires more than just a little knowledge of probability theory. At various points in his analysis, he modestly declares the math to be really nothing more than "elementary calculus" requiring only "routine evaluations" that are "easily calculated." I am sure he was quite serious about that when he wrote, but I also do think matters *are* just a bit more sophisticated than he claimed. After all, what is *your* reaction to his final expression for the kill probability, as a function of R and D?:

$$P_{kill}=\varphi(D+R)-\varphi(D-R)+\frac{R^2}{D^2}[\varphi(D+R)-\varphi(R-D)]$$
$$+\frac{1}{D^2\sqrt{2\pi}}[(D^2-RD+R^2-1)e^{-(R+D)^2/2}$$
$$-(D^2+RD+R^2-1)e^{-(D-R)^2/2}]$$

where

$$\varphi(x)=\int_{-\infty}^{x}\frac{1}{\sqrt{2\pi}}e^{-t^2/2}dt.$$

Mine is "Wow!"[7]

Professor Guenther evaluated his P_{kill} equation for numerous values of D and R (more details to come), but suppose we don't know the P_{kill} equation. That is, suppose you had been assigned the task of coming up with values for P_{kill} for various values of D and R but, alas, you don't have the required probability knowledge to duplicate Professor Guenther's analysis. Well, with a computer code like **ASW.m** you could still do it! Here's the code (notice how *short* it is) and a walk-through of how it works. The code simulates ten million torpedo attacks.

```
%ASW.m
01  D=0.5;R=3.5;kill=0;RS=R*R;
02  for loop=1:10000000
03        stop=0;
04        while stop==0
05              tx=-1+2*rand;
06              ty=-1+2*rand;
07              tz=-1+2*rand;
08              if tx^2+ty^2+tz^2<1
09                    stop=1;
10              end
11        end
12        tx=D*tx;ty=D*ty;tz=D*tz;
13        wx=randn;wy=randn;wz=randn;
14        if (tx-wx)^2+(ty-wy)^2+(tz-wz)^2<RS
15              kill=kill+1;
16        end
17  end
18  kill/loop
```

Before each attack, Line 01 defines the values of D and R that will be used in each of those attacks, initializes the number of kills to date to zero, and calculates, just *once*, the variable RS as R^2 (instead of recalculating R^2 every time it is needed). Lines 02 and 17 define the loop that executes the ten million attacks. The location of the hostile submarine, when detected, is taken to be the origin of a three-dimensional rectangular coordinate system (and the center of a sphere of radius D). To find the location in that sphere of the submarine when the torpedo warhead explodes, the code uses the basic idea developed in **wifi.m**. That is, to find a random point inside a circle, we embedded the circle in a square, generated random points over the square, and kept only those points that were also inside the circle. In **ASW.m** we embed a *unit* radius sphere in a cube, generate random points over the cube, and keep only those points that are also inside the sphere. This is done in Lines 03

TABLE 4.3.1. P_{kill} as Determined by **ASW.m**

$D \backslash R$	0.5	1.0	1.5	2.0	2.5	3.0	3.5
0	0.03	0.20	0.48	0.73	0.90	0.97	>0.99
0.5	0.03	0.19	0.46	0.72	0.89	0.97	>0.99
1.0	0.02	0.16	0.40	0.66	0.84	0.94	0.98
1.5	0.02	0.12	0.32	0.56	0.77	0.91	0.97
2.0	0.01	0.08	0.24	0.45	0.67	0.83	0.93
2.5	0.01	0.05	0.17	0.34	0.55	0.74	0.87
3.0	<0.01	0.03	0.11	0.25	0.42	0.62	0.78
3.5	<0.01	0.02	0.08	0.17	0.32	0.49	0.67

The unit of distance for both R and D is the standard deviation of the normally distributed, zero-mean aiming error.

through 11 (the calculation of *tx*, *ty*, and *tz*, where *t* stands for *target*). Line 12 then scales the final coordinates of the hostile submarine to a sphere of radius D (the final calculation of *tx*, *ty*, and *tz*). Line 13 determines where the warhead explodes (remember, the torpedo is aimed at the detection location of the hostile submarine—the origin of our coordinate system—and MATLAB's *randn* command generates normally distributed, zero-mean, unit standard deviation values, which are just what we need to model the aiming errors that determine where the warhead explodes). Lines 14, 15, and 16 determine if the target is within distance R of the warhead detonation and, if so, the code increments the variable *kill* by one. Line 18 gives us the fraction of the ten million torpedo attacks that resulted in a kill.

When run for various values of D and R, **ASW.m** produced the values for P_{kill} given in Table 4.3.1. Those values from the computer code, *without exception, are identical* to the values *computed* by Professor Guenther using his theoretical P_{kill} equation.

As a final comment, I have used the values for D and R that Professor Guenther used, simply to check that the **ASW.m** code is, indeed, properly modeling this problem. I have no idea where he got those values, and reasonable changes in the assumptions in the problem can result in vastly different values for R, in particular.

For example, in his paper Professor Guenther remarks, in passing, that the weapon launched at the hostile submarine might be nuclear in nature. Such a weapon offers a HUGE step-up in lethality. The warhead of the Mk-48 torpedo (see note 5) is several hundred *pounds* of conventional high explosives, while actual underwater detonations of nuclear devices have *started* at 30 kilo*tons* (twice the size of the atomic bomb dropped on Hiroshima, Japan, in 1945)—that is, at the energy equivalent of 30,000 metric tons (66 *million* pounds) of high explosive. Photos of such explosions make it easy to believe that no submarine within one thousand meters could possibly survive.[8]

5

A Problem in Electric Circuit Theory

5.1 The Mathematical Physics of Resistor Circuits

Before we start the core of this chapter (the longest in the book), consider a physical situation that you almost certainly will think couldn't possibly have anything to do with electrical circuits.

But it does.

Imagine a large water reservoir that services the needs of a nearby city. It is connected to the city by a vast number of pipes of various sizes that form some arbitrarily complicated network. The reservoir is at a higher elevation than the city, so water flows to the city through the pipe network because of gravity. Each individual pipe has a shut-off valve that can be set from fully open to fully closed. Suppose each pipe valve is set to half open; then clearly some volume of water will flow to the city each second (the so-called *flow rate*). Suppose further you arbitrarily select one of the pipes and fully *close* its valve. What do you think happens to the flow rate?

Most people would say, "The flow rate *decreases*." It wouldn't make intuitive sense, after all, that *eliminating* one of the flow paths should cause the flow rate to remain unchanged, and certainly not to *increase*. Think about this as you continue to read, and we will return to it at the end of the chapter. *Now* we start.

Most academic physicists, mathematicians, and engineers, over the course of decades of teaching, collect certain favorite problems that are always available in their notes as both entertaining and (they hope) educational examples for students. (Also, in the interest of honesty, for personal use as emergency, last-minute backup lecture material!) The charm of some of these problems is that while they may *seem* elementary, perhaps even *borderline trivial* at first glance, they quickly reveal their underlying computational horror *if attacked by "routine" thinking*. This chapter contains four such problems, each progressively more difficult, all based on the simplest electrical circuit element: the resistor.

A *resistor* is a two-terminal electrical device that obeys *Ohm's law* (named after German experimenter Georg Ohm [1789–1854]), which is a pretty simple algebraic equation. Electrical *circuits* made from the connection of multiple resistors together (and, to keep things as simple as possible, energized by a single voltage source, like a one-volt battery[1]) obey two additional, equally simple laws named after German physicist Gustav Kirchhoff (1824–1887). Such circuits, given the values of all the resistors and a description of how they are interconnected, can always, in principle, be solved through the use of the Ohm and Kirchhoff laws; that is, we can calculate the currents and voltages at every point in the circuit.

From a purely mathematical point of view, then, resistor circuits are "uninteresting" because, for a mathematician, nothing is left to be said. For an electrical engineer, however, there is a *lot* more to say, because just writing down equations based on the Ohm and Kirchhoff laws is a *lot* easier than solving them. This can become dramatically clear long before you get to complicated circuits that are constructed using lots of resistors: you can run into misery-inducing computational difficulties with circuits made from just a handful of resistors. After showing you some examples of this frustrating situation, and the clever "tricks" that solve them, the chapter concludes by developing a Monte Carlo algorithm for solving *any* resistor circuit, along with a computer code that implements that

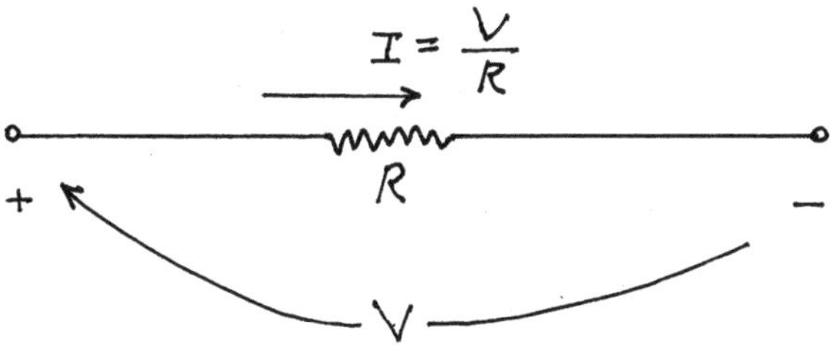

FIGURE 5.1.1. Ohm's law.

algorithm. This is the monster code of this book, using a *huge* number of random numbers (literally many tens of millions).

Before we start with specific circuits, here is a quick run through of the Ohm and Kirchhoff laws, just to be sure all readers start from the same point. Everything here is pretty basic (you can find it in the electrical discussions of any good high school physics text), and if it is old hat, feel free to skip ahead a few paragraphs.

> *Ohm's law* (Figure 5.1.1): If a voltage *difference* (or *drop*) of V volts[2] exists across a resistor of R ohms, then an electrical current (the motion of negative electrical charges called *electrons*[3]) of $I = \frac{V}{R}$ amperes[4] exists in the resistor, flowing from the end of the resistor at the higher voltage to the other end at the lower voltage. The *drop* is the voltage at the + end minus the voltage at the − end.

> *Kirchhoff's Current Law* (Figure 5.1.2): KCL states the sum of the currents flowing into any point in a circuit equals the sum of the currents flowing out of that point. This is often alternatively stated as the sum of *all* the currents *into* (*out of*) any point in a circuit is zero. This is a statement of the conservation of electrical charge: there is no charge accumulation at a node (like charges repel). Note: At least one of the currents in Figure 5.1.2 must be negative (a current that is actually flowing *away* from the central point).

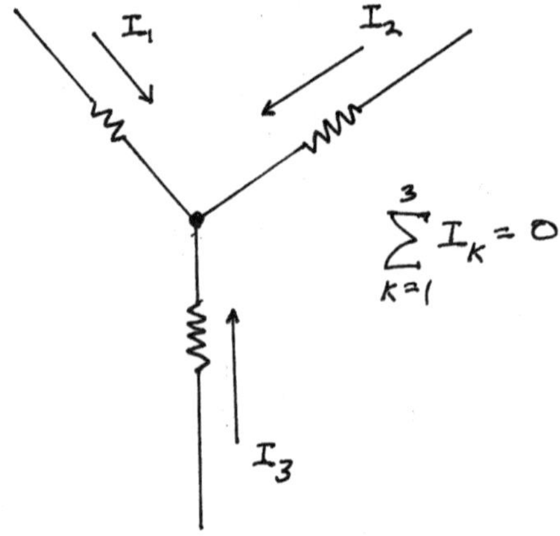

FIGURE 5.1.2. Kirchhoff's Current Law (KCL).

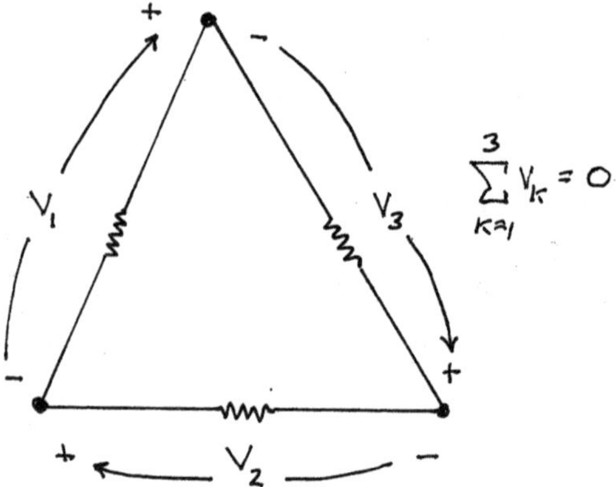

FIGURE 5.1.3. Kirchhoff's Voltage Law (KVL). The actual polarity of at least one of the voltages in the loop must be the opposite of what is shown.

Kirchhoff's Voltage Law (Figure 5.1.3): KVL states the sum of the voltage drops around any closed loop is zero. This is a statement of the conservation of energy.

In elementary introductions to electrical circuits, an analogy is often made between electricity flowing in circuits made of resistors and water flowing through a network of pipes. In that analogy, the following are pairs: water pressure and voltage, water and electrical charge, water flow rate and electrical current, and the shut-off valve settings of the pipes and the values of resistors (the higher the resistance, the more a valve is shut off). Recall this analogy at the end of the chapter when we return to the opening question about the water reservoir.

Here follows a description of four circuits (in order of increasing difficulty). First, you will see a slick way (that is, a trick) to get a solution for the first three without having to solve a large number of equations (the fourth circuit will be just a bit more complicated). Then we will finish with the development of a general-purpose computer code that solves all four circuits (and *vastly* more complicated circuits as well), a particularly helpful code when you have no more tricks to pull out of your hat. The theoretical solutions found in the next three sections verify the correctness of the code.

5.2 The Resistor Cube

So, to start, consider the *3-dimensional resistor cube* of Figure 5.2.1, with all the resistors equal to R ohms. The one-volt battery is connected to the *nodes* (a point where two or more resistors connect) labeled ⓪ and ⑦.[5] Finally, let's agree to define the negative terminal of the battery to be at zero volts (that is, to serve as the reference to which all other voltages will be measured). So, $v_0 = 1$ and $v_7 = 0$. Our problem is to determine the values of the other six voltages, v_1 through v_6.

The direct way to approach this problem is, of course, with the laws of Ohm and Kirchhoff. That is, let's write the KCL equations for nodes ① through ⑥. Note *carefully* that we do *not* write a KCL equation for either node ⓪ or node ⑦. Do you see why we exclude those two nodes? It's not because KCL doesn't hold there (it does), but rather because we would have to include the battery current at those nodes, which we don't know yet. After we have

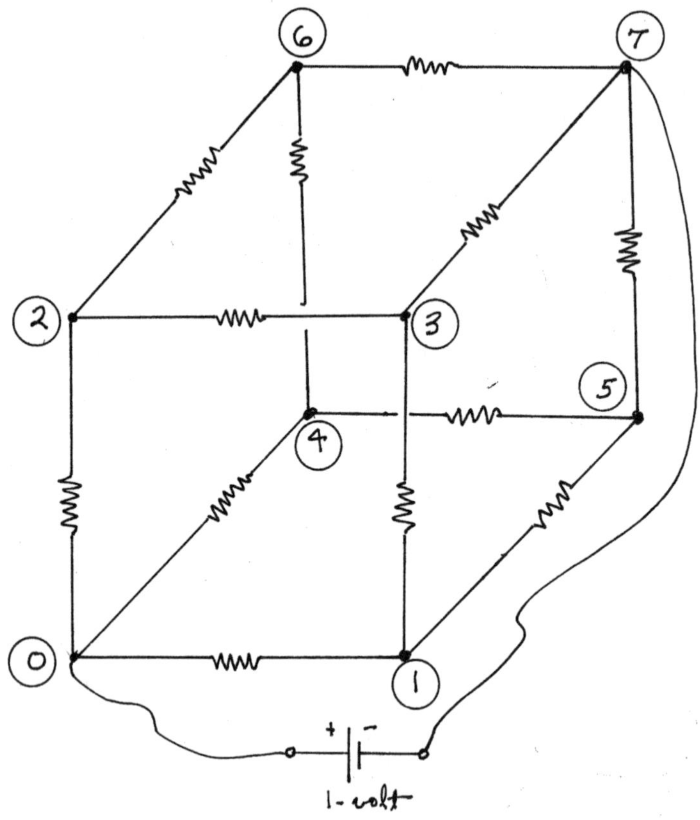

FIGURE 5.2.1. The 3D-resistor cube.

solved all the node voltages, we can *then* use the results to determine the battery current.[6] So, we have

$$\frac{1-v_1}{R}+\frac{v_3-v_1}{R}+\frac{v_5-v_1}{R}=0 \text{ (KCL at node ①)}$$

$$\frac{1-v_2}{R}+\frac{v_3-v_2}{R}+\frac{v_6-v_2}{R}=0 \text{ (KCL at node ②)}$$

$$\frac{v_2-v_3}{R}+\frac{v_1-v_3}{R}+\frac{0-v_3}{R}=0 \text{ (KCL at node ③)}$$

$$\frac{1-v_4}{R}+\frac{v_5-v_4}{R}+\frac{v_6-v_4}{R}=0 \text{ (KCL at node ④)}$$

$$\frac{v_1-v_5}{R}+\frac{v_4-v_5}{R}+\frac{0-v_5}{R}=0 \text{ (KCL at node ⑤)}$$

$$\frac{v_2-v_6}{R}+\frac{v_4-v_6}{R}+\frac{0-v_6}{R}=0 \text{ (KCL at node ⑥)}$$

or, noticing the *R*s cancel (and so the specific value of *R* is irrelevant),

$$\begin{gathered}1+v_3+v_5-3v_1=0\\ 1+v_3+v_6-3v_2=0\\ v_2+v_1-3v_3=0\\ 1+v_5+v_6-3v_4=0\\ v_1+v_4-3v_5=0\\ v_2+v_4-3v_6=0.\end{gathered}$$

Six equations in six unknowns. Solvable, yes, but still, *what a mess*! Or it is until we notice that, by the symmetry of the cube (*because all the resistors are equal*), we have $v_1=v_2=v_4=x$ and $v_3=v_5=v_6=y$ and the six equations become

(a) $1+2y-3x=0$
(b) $1+2y-3x=0$
(c) $2x-3y=0$
(d) $1+2y-3x=0$
(e) $2x-3y=0$
(f) $2x-3y=0$.

Equations (a), (b), and (d) are identical, as are equations (c), (e), and (f). So, all we really have are the *two* equations $1+2y-3x=0$ and $2x-3y=0$, which are easily solved to give

$$x=v_1=v_2=v_4=\frac{3}{5}\text{ volt (0.6000 volt)} \qquad (5.2.1)$$

and

$$y=v_3=v_5=v_6=\frac{2}{5}\text{ volt}\,(0.4000\text{ volt}). \tag{5.2.2}$$

This is nice, yes, but don't forget: *it's all because the resistors were taken to be equal.* If the resistors had different values, then the cube would lose its symmetry and we *would* have had to solve six equations in six unknowns. Upon developing our general algorithm and its implementation as a computer code, the loss of symmetry will be of no consequence. We will use (5.2.1) and (5.2.2) as test cases for the code.

Finally, to fulfill the promise I made in note 6, we see that the battery current is

$$\frac{v_0-v_1}{R}+\frac{v_0-v_4}{R}+\frac{v_0-v_2}{R}=\frac{1-\frac{3}{5}}{R}+\frac{1-\frac{3}{5}}{R}+\frac{1-\frac{3}{5}}{R}$$
$$=\frac{6}{5R}\text{ amperes.}$$

This result is the solution to a classic problem in electric circuit theory, one that physics and electrical engineering professors like to spring on students (this question is not really of direct relevance for this chapter, but it is just too sweet to let pass without mention). It is to calculate the equivalent resistance (R_{eq}) of a resistor cube made of one-ohm resistors when measured between the two nodes that are farthest apart (like nodes ⓪ and ⑦). That is, what is the so-called *body-diagonal resistance*? With $R=1$, we see that a one-volt drop across nodes ⓪ and ⑦ produces a current of $\frac{6}{5}$ ampere, and so Ohm's law states $1=\frac{6}{5}R_{eq}$ or, $R_{eq}=\frac{5}{6}$ ohm. How to make this calculation, using the usual rules for how resistors in series and parallel combine, isn't obvious.

5.3 The Pinwheel Circuit

For a second example of how a mere handful of resistors can form a circuit of considerable challenge, consider the *pinwheel* circuit of Figure 5.3.1, with all twelve of the resistors equal to one ohm. We

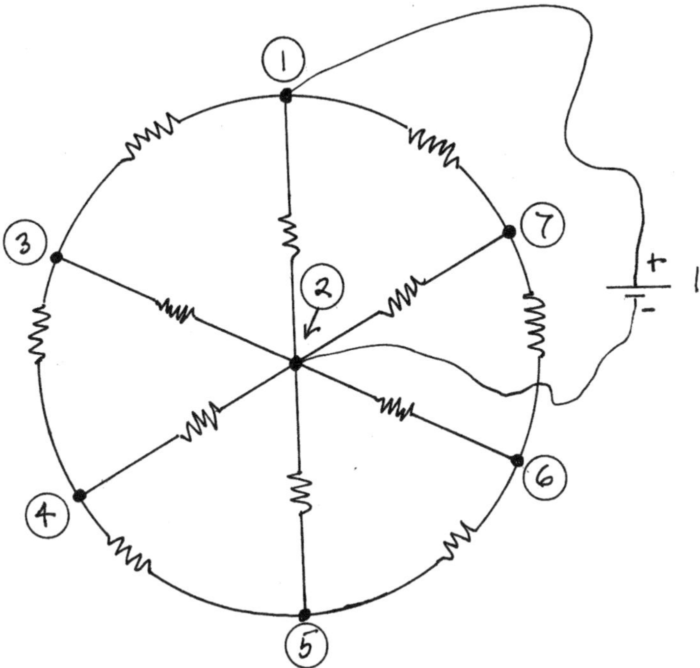

FIGURE 5.3.1. A pinwheel resistor circuit.

know $v_1 = 1$ and $v_2 = 0$, and our problem is to compute the voltages of the other nodes, v_3 through v_7. Writing the KCL equations for the five nodes labeled ③ through ⑦ (remember to stay away from the nodes connected to the battery), we get (remember, too, that each resistor is one ohm) the following five equations in five variables:

$$(1-v_3)+(0-v_3)+(v_4-v_3)=0 \quad \text{(KCL at node ③)}$$
$$(v_3-v_4)+(v_5-v_4)+(0-v_4)=0 \quad \text{(KCL at node ④)}$$
$$(v_4-v_5)+(v_6-v_5)+(0-v_5)=0 \quad \text{(KCL at node ⑤)}$$
$$(v_5-v_6)+(v_7-v_6)+(0-v_6)=0 \quad \text{(KCL at node ⑥)}$$
$$(v_6-v_7)+(1-v_7)+(0-v_7)=0 \quad \text{(KCL at node ⑦)}$$

or,

$$1+v_4-3v_3=0,$$
$$v_3+v_5-3v_4=0,$$

$$v_4+v_6-3v_5=0,$$
$$v_5+v_7-3v_6=0,$$
$$1+v_6-3v_7=0.$$

Now, by the symmetry of Figure 5.3.1, $v_3=v_7=x$, $v_4=v_6=y$, and $v_5=z$. Thus,

(a) $1+y-3x=0$
(b) $x+z-3y=0$
(c) $2y-3z=0$
(d) $z+x-3y=0$
(e) $1+y-3x=0.$

Since (a) and (e) are identical and (b) and (d) are identical, then all we have are *three* equations in three variables (remember, this is for the special case where *all the resistors are one ohm*):

$$1+y-3x=0$$
$$x+z-3y=0$$
$$2y-3z=0.$$

These three simultaneous equations are easy to solve (you should do that), with the result

$$x=v_3=v_7=\frac{7}{18}=0.3889 \text{ volt} \tag{5.3.1}$$

$$y=v_4=v_6=\frac{1}{6}=0.1667 \text{ volt} \tag{5.3.2}$$

$$z=v_5=\frac{1}{9}=0.1111 \text{ volt.} \tag{5.3.3}$$

5.4 The Ladder and Lattice Circuits

For the third difficult circuit example (with a new trick), consider Figure 5.4.1, a circuit that electrical engineers (who are, mostly, poets at heart) call a *ladder circuit*. The resistor values r_1 through r_9 are *arbitrary*. They might all be equal, but they don't have to be.

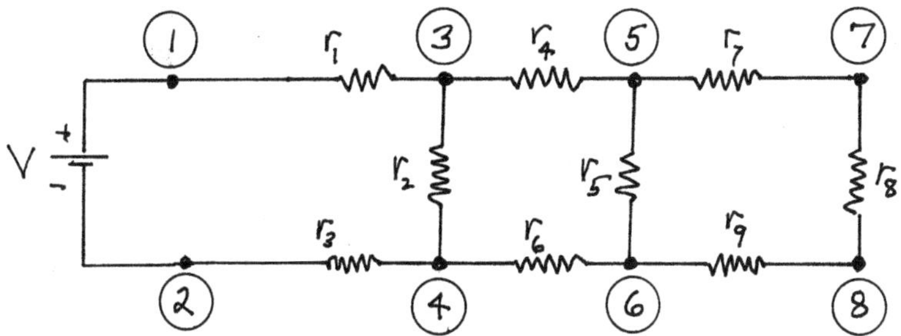

FIGURE 5.4.1. A ladder circuit.

Our problem again is to determine the voltages at nodes ③ through ⑧ when the circuit is energized by a 1-volt battery (I will soon explain why the voltage source is shown as V and not as 1). Again, we can write the KCL equations at nodes ③ through ⑧, a total of six equations in the six unknowns v_3 through v_8. (Remember, we know from the start that $v_1 = V$ and $v_2 = 0$.) The new difficulty is there is no obvious symmetry in this circuit's node voltages that allows us to reduce the number of equations. It seems that now we really *are* going to have to solve six simultaneous equations in six unknowns. Either that, or we need to find a new trick.

And here it is! To make the trick's explanation as transparent as possible, suppose all the resistors are one ohm. You will see at the end of the following analysis that this assumption in no way limits the success of the trick, but simply makes the arithmetic *so* easy we can do it in our heads. The trick is to solve the circuit *backwards*. That is, to start at the right-most part of the circuit and work to the left, toward the voltage source.

So, suppose we have adjusted the battery voltage (see note 1 again) to be whatever V needs to be to result in a 1 ampere current in r_8 flowing downward (and so the current in r_7 is 1 ampere to the right, and 1 ampere to the left in r_9, as those three resistors are in series). The total voltage drop across r_7, r_8, and r_9 is therefore 3 volts, which is the voltage drop across r_5. So, the current in r_5 is (from

Ohm's law) 3 amperes, flowing downward. Applying KCL at node ⑤ finds 4 amperes must be flowing to the right in r_4, and applying KCL at node ⑥ finds 4 amperes must be in r_6, flowing to the left. Continuing, there is a 4-volt drop across r_4, and a 4-volt drop across r_6. With the 3-volt drop across r_5, we have a total voltage drop from node ③ to node ④ of 11 volts, which means 11 amperes in r_2 are flowing downward. Applying KCL at node ③, we see there is a current of 15 amperes flowing to the right in r_1, and applying KCL at node ④ means 15 amperes in r_3 are flowing to the left. Thus, the total voltage drop from node ① to node ② is 41 volts. That is, $V = v_1 - v_2 = 41$ volts is required to get 1 ampere in r_8.

With our knowledge of the currents in each of the resistors, we can use Ohm's law to calculate the voltage drop across each resistor, giving us the node voltages. So, with $V = 41$ volts,

$$v_3 = 41 - 15 = 26 \text{ volts}, \; v_4 = v_3 - 11 = 26 - 11 = 15 \text{ volts},$$
$$v_5 = v_3 - 4 = 26 - 4 = 22 \text{ volts}, \; v_6 = v_5 - 3 = 22 - 3 = 19 \text{ volts},$$
$$v_7 = v_5 - 1 = 22 - 1 = 21 \text{ volts}, \; v_8 = v_7 - 1 = 21 - 1 = 20 \text{ volts}.$$

Now, just one last step.

Our problem was to determine the node voltages *not* for $V = 41$ volts, but rather for $V = 1$ volt (you will see in the next section *why* this apparent obsession with $V = 1$ volt). To address this issue, we simply *scale* the above voltages by a factor of 41 to get (see the following box for a discussion on *why* scaling works)

$$v_3 = \frac{26}{41} = 0.6341 \text{ volt} \tag{5.4.1}$$

$$v_4 = \frac{15}{41} = 0.3658 \text{ volt} \tag{5.4.2}$$

$$v_5 = \frac{22}{41} = 0.5366 \text{ volt} \tag{5.4.3}$$

$$v_6 = \frac{19}{41} = 0.4634 \text{ volt} \tag{5.4.4}$$

We can replace Figure 5.4.1 with Figure 5.4.2, where the battery is placed outside a box. (This is the famous "black box" engineers and physicists often invoke when the details of its interior are not important.) The box contains the resistor network (which could be *any* arrangement of resistors, not just the ladder of Figure 5.4.1). As far as the battery is concerned, it simply "sees" some equivalent resistance R_{eq} as it "looks into" the box. We used this idea when discussing the body-diagonal resistance of the symmetrical 3D-resistor cube in section 5.2.

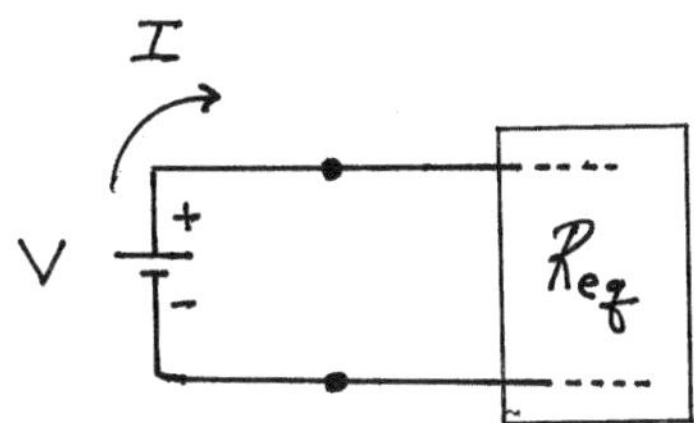

FIGURE 5.4.2. The generalization of Figure 5.4.1.

The battery current is, from Ohm's law, $I = \frac{V}{R_{eq}}$. The battery is sending electrical energy to the box at a rate given by the power (measured in watts, a well-known unit named after Scottish engineer James Watt ([1736–1819]). The battery power is $VI = \frac{V^2}{R_{eq}}$, where V is in volts and I is in amperes. The battery electrical energy is totally converted to heat energy, with each individual resistor r_j dissipating energy at the rate of $\frac{e_j^2}{r_j}$, where $1 \leq j \leq n$ with n equal to the number of resistors in the box (as mentioned, the details of *how* the resistors are connected are unimportant) and e_j is the voltage *drop* across r_j. By conservation of energy, the power supplied by the battery is the total power dissipated by the resistors, and so $\frac{V^2}{R_{eq}} = \sum_{j=1}^{n} \frac{e_j^2}{r_j}$. Now, suppose we change V to $\frac{V}{k}$ where k is some constant. Then we have the power supplied by the battery as $\frac{\left(\frac{V}{k}\right)^2}{R_{eq}}$ and so if e_j' is the new voltage drop across the jth resistor, we must have $\frac{V^2}{k^2 R_{eq}} = \sum_{j=1}^{n} \frac{e_j'^2}{r_j} = \frac{1}{k^2} \sum_{j=1}^{n} \frac{e_j^2}{r_j} = \sum_{j=1}^{n} \frac{\left(\frac{e_j}{k}\right)^2}{r_j}$. Since this has to hold for any choice of values for r_j, we conclude that $e_j' = \frac{e_j}{k}$, from which it follows that the individual node voltages scale by k, too.

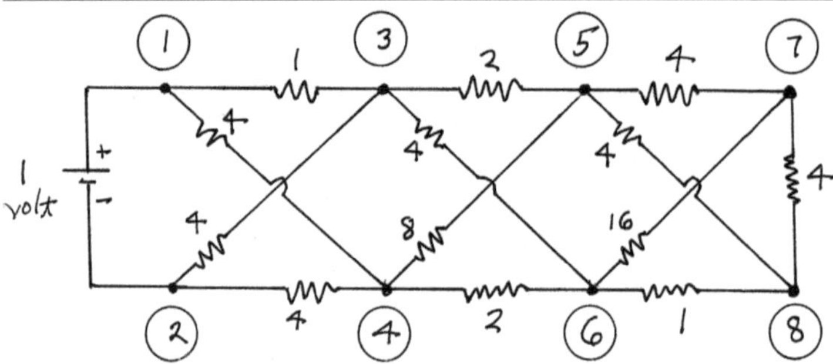

FIGURE 5.4.3. A lattice circuit of resistors.

$$v_7 = \frac{21}{41} = 0.5122 \text{ volt} \tag{5.4.5}$$

$$v_8 = \frac{20}{41} = 0.4878 \text{ volt.} \tag{5.4.6}$$

With this—the last of the first three circuits we will use to check the computer code to be developed in the next two sections—let me address a concern of mine: you might have gotten the impression that there will always be a trick lurking about in the background, for just about any circuit we study, to save the day. Alas, not so. For example, as a final theoretical analysis, consider the circuit in Figure 5.4.3, which bears a strong resemblance to the ladder but is sufficiently different that poetic electrical engineers give it another name: a *lattice* (the values of the resistors are given by the numbers next to each resistor).

The "working backwards" trick doesn't work here, and there is no symmetry-based trick either. Of course, KCL always works, and so we will do that and then see what a mathematician might say about it all. Since $v_1 = 1$ and $v_2 = 0$, we have

$$\frac{1-v_3}{1} + \frac{0-v_3}{4} + \frac{v_5-v_3}{2} + \frac{v_6-v_3}{4} = 0 \quad \text{(KCL at node ③)}$$

$$\frac{0-v_4}{4} + \frac{1-v_4}{4} + \frac{v_5-v_4}{8} + \frac{v_6-v_4}{2} = 0 \quad \text{(KCL at node ④)}$$

$$\frac{v_3 - v_5}{2} + \frac{v_4 - v_5}{8} + \frac{v_7 - v_5}{4} + \frac{v_8 - v_5}{4} = 0 \quad \text{(KCL at node ⑤)}$$

$$\frac{v_4 - v_6}{2} + \frac{v_3 - v_6}{4} + \frac{v_7 - v_6}{16} + \frac{v_8 - v_6}{1} = 0 \quad \text{(KCL at node ⑥)}$$

$$\frac{v_5 - v_7}{4} + \frac{v_6 - v_7}{16} + \frac{v_8 - v_7}{4} = 0 \quad \text{(KCL at node ⑦)}$$

$$\frac{v_5 - v_8}{4} + \frac{v_6 - v_8}{1} + \frac{v_7 - v_8}{4} = 0 \quad \text{(KCL at node ⑧).}$$

Clearing out the fractions, these equations become

$$\begin{gathered}
4 - 4v_3 - v_3 + 2v_5 - 2v_3 + v_6 - v_3 = 0 \\
-2v_4 + 2 - 2v_4 + v_5 - v_4 + 4v_6 - 4v_4 = 0 \\
4v_3 - 4v_5 + v_4 - v_5 + 2v_7 - 2v_5 + 2v_8 - 2v_5 = 0 \\
8v_4 - 8v_6 + 4v_3 - 4v_6 + v_7 - v_6 + 16v_8 - 16v_6 = 0 \\
4v_5 - 4v_7 + v_6 - v_7 + 4v_8 - 4v_7 = 0 \\
v_5 - v_8 + 4v_6 - 4v_8 + v_7 - v_8 = 0
\end{gathered}$$

or, combining variable terms on the left and putting constant terms on the right, we have

$$\begin{gathered}
-8v_3 + 2v_5 + v_6 = -4 \\
-9v_4 + v_5 + 4v_6 = -2 \\
4v_3 + v_4 - 9v_5 + 2v_7 + 2v_8 = 0 \\
4v_3 + 8v_4 - 29v_6 + v_7 + 16v_8 = 0 \\
4v_5 + v_6 - 9v_7 + 4v_8 = 0 \\
v_5 + 4v_6 + v_7 - 6v_8 = 0.
\end{gathered}$$

Looking at these six equations, it seems impossible to avoid the conclusion that we face a *challenging* problem in computing the node voltages v_3 through v_8! A mathematician, however, would be unmoved by our horror and simply say, "Use Cramer's rule." This refers to the method published in 1750 by Swiss mathematician Gabriel Cramer (1704–1752) for solving n simultaneous equations in n variables with the use of determinants. Often taught in high school algebra, Cramer's rule requires the calculation of $n+1$, n-by-n determinants. Thus, our lattice circuit KCL equations, with $n=6$, require us to calculate seven 6-by-6 determinants (a nontrivial task!) as follows. We start by writing each of the six equations such that *all* the node voltage variables on the left of the equality are present (if a variable does not actually occur in an equation, include it anyway with a zero coefficient). Doing that, the KCL equations for Figure 5.4.3 become

$$\begin{gathered}
-8v_3+0v_4+2v_5+1v_6+0v_7+0v_8=-4\\
0v_3-9v_4+1v_5+4v_6+0v_7+0v_8=-2\\
4v_3+1v_4-9v_5+0v_6+2v_7+2v_8=0\\
4v_3+8v_4+0v_5-29v_6+1v_7+16v_8=0\\
0v_3+0v_4+4v_5+1v_6-9v_7+4v_8=0\\
0v_3+0v_4+1v_5+4v_6+1v_7-6v_8=0.
\end{gathered}$$

Writing these equations in matrix form (this math is probably *not* taught in most high school algebra classes), these last equations become[7]

$$\begin{bmatrix} -8 & 0 & 2 & 1 & 0 & 0 \\ 0 & -9 & 1 & 4 & 0 & 0 \\ 4 & 1 & -9 & 0 & 2 & 2 \\ 4 & 8 & 0 & -29 & 1 & 16 \\ 0 & 0 & 4 & 1 & -9 & 4 \\ 0 & 0 & 1 & 4 & 1 & -6 \end{bmatrix} \begin{bmatrix} v_3 \\ v_4 \\ v_5 \\ v_6 \\ v_7 \\ v_8 \end{bmatrix} = \begin{bmatrix} -4 \\ -2 \\ 0 \\ 0 \\ 0 \\ 0 \end{bmatrix},$$

an expression typically written by electrical engineers as

$$Av=b \tag{5.4.7}$$

where A is the 6-by-6 matrix, v is a column vector of the node voltage variables, and b is the column vector of constants on the right of the equality in (5.4.7). Now, to find the determinants for use in Cramer's rule, we proceed next by doing. . . .

At this point another mathematician, who just happens to be passing by, interrrupts to say, "Holy cow, *stop*, forget Cramer's rule; just multiply through (5.4.7) with the inverse matrix A^{-1} to directly solve for $\boldsymbol{v}$." That is, write $A^{-1}Av = A^{-1}b$ and since (by the very definition of an inverse matrix) $A^{-1}A = I$ (I is the 6-by-6 identity matrix), and since $Iv = v$, then the answer is $v = A^{-1}b$. Well, all this is just great, except for the possibility you don't know anything about matrix theory or, even if you do, do you *really* want to compute the inverse of a 6-by-6 matrix? Probably not (unless you're a glutton for grubby arithmetic). But, fortunately, it's all duck soup for MATLAB, which has the wonderful command inv(A) that computes A^{-1} at the touch of a button. Not to keep you in suspense, doing all this produces the result[8]

$$v = \begin{bmatrix} 0.7608 \\ 0.5980 \\ 0.7090 \\ 0.6683 \\ 0.6911 \\ 0.6789 \end{bmatrix}. \tag{5.4.8}$$

With this, the last of what will serve as the test cases to verify the computer code (to be created in section 5.6) that can handle *any* resistor circuit, we are ready to develop the underlying mathematical theory of that code.

5.5 A Random Walk through a Circuit

We start by considering an arbitrary connection of resistors, as shown in Figure 5.5.1, with any pair of nodes either not connected or, if connected, the connection is via a single resistor. The figure shows, for example, that nodes x and y are connected by resistor r_{xy} and that the current in r_{xy} is $I_{x\to y}$, flowing from x to y. Two particular nodes, A and B, are connected to the terminals of a one-volt battery.[9] If V_x denotes the voltage at node x, then $V_A = 1$ and $V_B = 0$. The voltage drop across r_{xy} is $V_x - V_y = I_{x\to y}r_{xy}$. Our problem, as it has been throughout this chapter, is to determine the voltages of all the other nodes in the circuit.

Let's continue to consider the two arbitrary nodes x and y in Figure 5.5.1 that are connected by the resistor r_{xy}. There may possibly be multiple nodes connected to node x, and so node y is just one of perhaps many (in Figure 5.5.1 there are three nodes connected to node x, one of which is B). Let's write $N(x)$ as the set of all the nodes connected to node x and so, since y is a member of that set, the usual mathematical notation is written $y \in N(x)$. Note, carefully, that y is now a stand-in for *any* node connected to x. Writing KCL at node x, we have

$$\sum_{y\in N(x)} I_{x\to y} = 0 = \sum_{y\in N(x)} \frac{V_x - V_y}{r_{xy}}$$
$$= V_x \sum_{y\in N(x)} \frac{1}{r_{xy}} - \sum_{y\in N(x)} \frac{V_y}{r_{xy}}. \quad (5.5.1)$$

We next define

$$c_x = \sum_{y\in N(x)} \frac{1}{r_{xy}} \quad (5.5.2)$$

and so (5.5.1) becomes

$$V_x c_x - c_x \frac{1}{c_x} \sum_{y\in N(x)} \frac{V_y}{r_{xy}} = 0$$

or,

$$c_x \left(V_x - \sum_{y\in N(x)} \frac{V_y}{c_x r_{xy}} \right) = 0. \quad (5.5.3)$$

To see what (5.5.3) is telling us, we make this simple physical observation: $c_x \neq 0$. Why is that? Looking at (5.5.2), we see that the *only* way $c_x = 0$ is if *every* resistor between node x and the nodes x is connected to is *infinite* in value. But that is just a fancy way of

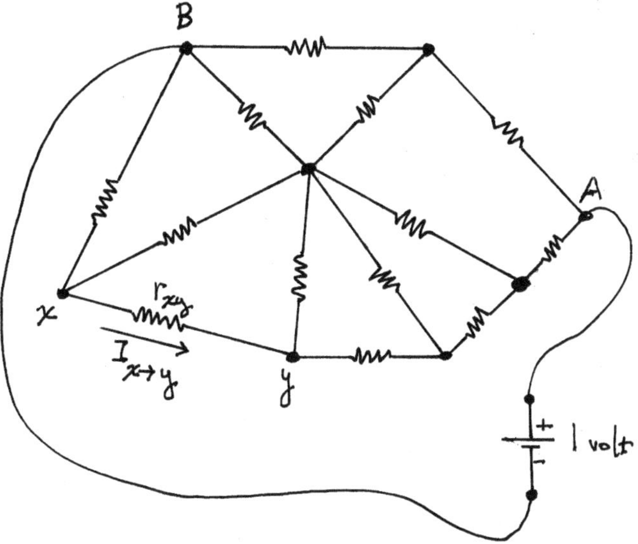

FIGURE 5.5.1. A "typical" arbitrary resistor network with a single voltage source.

saying that x is *not* really connected to *any* node, and so $c_x = 0$ would mean x is *not* present in the circuit! So, for any circuits where node x *is* present, $c_x \neq 0$. This conclusion means that the *other* factor on the right of (5.5.3) must be zero (at least one of those factors has to be zero for their product to be zero!). This gives us what will prove to be of crucial importance in developing the computer code for resistor circuits:

$$V_x = \sum_{y \in N(x)} \frac{V_y}{c_x r_{xy}}. \tag{5.5.4}$$

Now, a brief pause to wax a bit philosophical. Up to this point, we have been thinking like electrical engineers, using the laws of Ohm and Kirchhoff. From now on we will proceed with the viewpoint of a mathematician, for whom an electron moving through the wires of a resistor circuit is simply what is called a *random walker.*[10] That is, a mathematician views an electron moving through the wires of an electrical circuit, transitioning as it goes from its present node to the next node, as obeying the following "rule":

> If an electron arrives at node x, it then promptly moves on to one of the nodes connected to x, with the electron "deciding" to move to node $y \in N(x)$ with the *transition probability*
>
> $$p_{x\to y} = \frac{1}{c_x r_{xy}} \text{ for all } y \in N(x). \tag{5.5.5}$$

It is very important to understand that nothing demands that this rule flow from pure algebraic logic. It is simply a *definition* (remember, in mathematics we are free to define anything we wish, with the value of the definition being established *if* it results in something useful). The above definition is a *creative act* (that is, it isn't *a priori* "obvious"[11]) that does indeed lead to something useful.

Let's first establish that we can, indeed, treat $p_{x\to y}$ in (5.5.5) as a probability. That is, show that $p_{x\to y}$ is always a number between 0 and 1. This is actually pretty easy to do. First, since $r_{xy} > 0$ (all the resistors in our circuit are positive[12]), then from (5.5.2) we see that $p_{x\to y} > 0$. Second, if we sum over *all* the $p_{x\to y}$, we have

$$\sum\nolimits_{y\in N(x)} p_{x\to y} = \sum\nolimits_{y\in N(x)} \frac{1}{c_x r_{xy}} = \frac{1}{c_x} \sum\nolimits_{y\in N(x)} \frac{1}{r_{xy}} = \frac{1}{c_x}(c_x) = 1.$$

So, since each $p_{x\to y} > 0$, and since all the $p_{x\to y}$ sum to 1, then each individual $p_{x\to y}$ must be bounded from above by 1. Notice that since all the transition probabilities are greater than 1 and sum to 1, there is zero probability of a node $x \to$ node x transition. That is, when an electron arrives at node x, it promptly moves on to some *other* (connected) node and doesn't just sit at node x. This behavior automatically ensures that a random walking electron obeys KCL.

The walk continues until the electron reaches either node A or node B, with each possibility terminating the walk. This is certain to eventually happen but, because of the randomness of the node-to-node transitions, it isn't clear how long that may take. It is perfectly possible for a random walking electron to fall into a loop and simply bounce back and forth between a pair of connected nodes. I will return to this concern in the next section when we write our computer code.

We are almost done with our mathematical analysis; pay close attention to the next few lines as our central result will appear suddenly. Define P_x as the probability that an electron, if at node x, will in its subsequent random wanderings through the resistor circuit eventually reach node A (connected to the positive terminal of the one-volt battery in Figure 5.5.1) *before* it reaches node B (connected to the negative terminal of the one-volt battery). We can model this random wandering as a two-stage process: starting at node x, the electron first goes to some connected node y and then, from node y, the electron eventually reaches (with probability P_y) node A. That is, since y is any of the nodes in the set $N(x)$, we have, by summing over all possible y,

$$P_x = \sum\nolimits_{y \in N(x)} p_{x \to y} P_y. \tag{5.5.6}$$

We justify the multiplication of the transition probability $p_{x \to y}$ with the probability P_y by remembering that each node transition choice by an electron is independent of any other transition decision it makes. Putting (5.5.5) into (5.5.6), we have

$$P_x = \sum\nolimits_{y \in N(x)} \frac{P_y}{c_x r_{xy}}. \tag{5.5.7}$$

Did you just have a bright flash of insight? If not, then compare (5.5.7) with (5.5.4). *They are the same equation*, with the probabilities P_x in (5.5.7) and the voltages V_x in (5.5.4) playing analogous roles! Notice that the association of P_x with V_x results in two obviously true statements for the special cases where x is node A and where x is node B. If $x = \text{A}$ then the probability of reaching A before reaching B is 1 (the electron is *already* at A and so the walk over), and $V_A = 1$ (this is why the battery voltage has always been specified to be *one* volt). If $x = \text{B}$ then the walk is over and the probability of reaching A before B is obviously 0, and $V_B = 0$. If you have a circuit with a battery voltage different from 1, solve it for a voltage of 1 and then scale as discussed in the box of section 5.4.

If we can find P_x then we have also found V_x, and finding P_x is *easy* to do—with a computer. The next (and final) section shows how to do that.

5.6 A Computer Code for *Any* Resistor Circuit

As just shown, to find P_x (and so V_x) we launch a *lot* of random walks from node x (with x as any one of the circuit nodes other than A or B) and keep track of how many of those walks terminate on node A. How many is a "*lot* of random walks"? The more the better, and here we will use ten million. So, for the 3D-resistor cube circuit of Figure 5.2.1, for example, with eight nodes (counting A and B, labeled as node ⓪ and node ⑦, respectively, in the figure), we will launch a total of 60 million walks.

Here is our first challenge: How do we tell the computer the connection details of the circuit our code is going to random walk on? That is, which nodes are connected, and what is the connecting resistance for each pair of connected nodes? We will start with the following rule for labeling the nodes.

> The positive terminal of the battery will *always* connect to node ①, and the negative terminal of the battery will *always* connect to node ②. All the other circuit nodes will be arbitrarily numbered from ③. There is no node ⓪. This rule is a result of MATLAB not allowing a *zero* index for the elements of a vector (an index must be a *positive* integer).

Looking back at the pinwheel circuit (Figure 5.3.1), the ladder circuit (Figure 5.4.1), and the lattice circuit (Figure 5.4.3), this *is* how the circuit nodes there were labeled. The 3D-resistor cube in Figure 5.2.1, however, used a different labeling of the nodes and

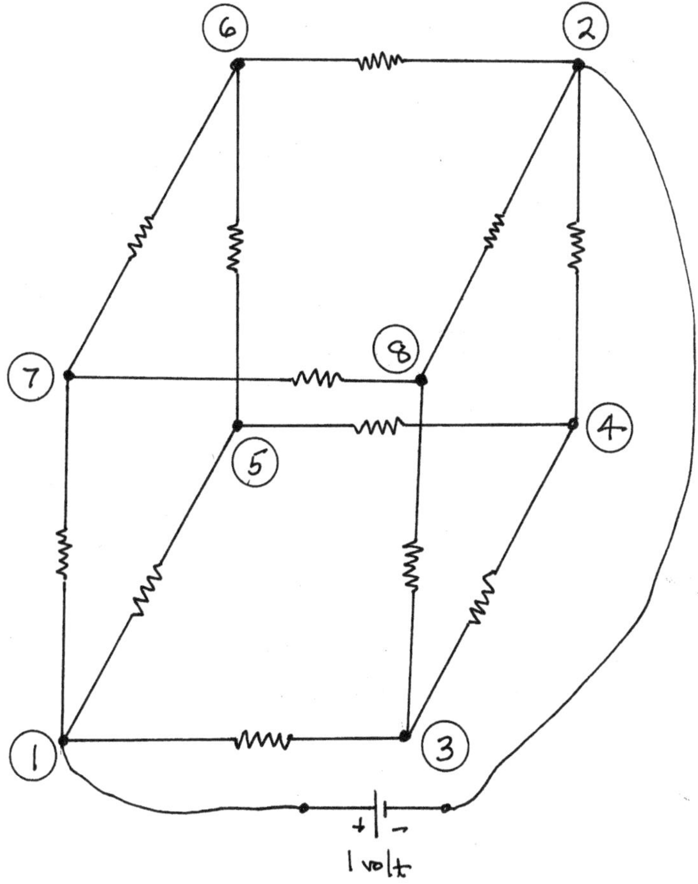

FIGURE 5.6.1. 3D-resistor cube with relabeled nodes.

so, to bring that circuit into compliance with our new labeling rule, we will use Figure 5.6.1 instead.

Recalling the theoretical values for the node voltages derived in (5.2.1) and (5.2.2), we have

$$v_3 = v_5 = v_7 = 0.6, \quad \text{and} \quad v_4 = v_6 = v_8 = 0.4. \qquad (5.6.1)$$

Now that the circuit nodes are labeled, we will tell our code *how* those nodes are connected by constructing a connection matrix

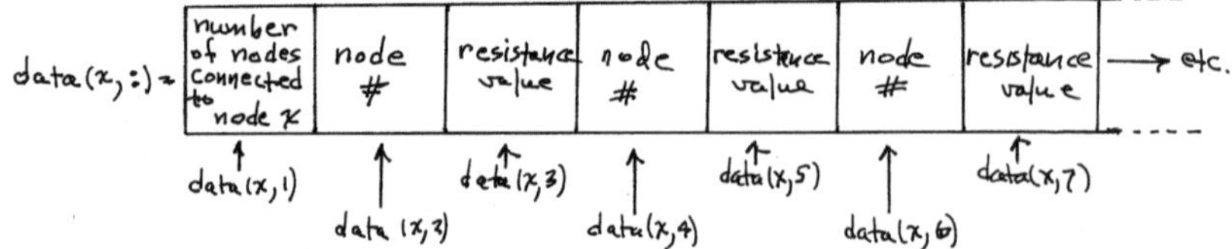

FIGURE 5.6.2. The xth row (for node x) in the connection matrix *data*. The colon in *data*(x,:) is MATLAB's way of specifying the entire row, and not just referencing a specific element in the row.

called *data,* in which each row of that matrix is uniquely associated with a specific node (the xth row goes with the xth node). The structure of the xth row is shown in Figure 5.6.2. The length of the row is $1+2*data(x, 1)$, where the first term of 1 accounts for the first element in the row (holding an integer equal to the number of nodes connected to node x (what we called the y nodes in the text). The second term is there because each such connected node requires two elements of the *data* row (one for the connected node number, and one for the connecting resistance). Since *data* is a matrix, MATLAB requires each row to have the same length, but that will happen only if each node is connected to the same number of nodes, which is not always the case. For example, in the lattice circuit of Figure 5.4.3, node ③ connects to four nodes, while node ⑦ connects to only three nodes. For such circuits, all the shorter rows in *data* must be "padded out" with zeros to make *all* row lengths equal. In fact, our code never actually looks at those padding zeros.

We are now ready to create the individual rows of *data* for the circuit we want to study. The first two rows of *data,* for node ① and node ②, are actually of no interest, because the function of a row is to tell the random walking electron which node to go to next, and there is no "next node" if the electron is at node ① or at node ②! So, we need only specify the rows of *data* starting with row 3. For the 3D-resistor cube of Figure 5.6.1, we have the description

```
numberofnodes=8;
data(3,:)=[3 1 1 4 1 8 1];
```

data(4,:)=[3 2 1 3 1 5 1];
data(5,:)=[3 1 1 4 1 6 1];
data(6,:)=[3 2 1 5 1 7 1];
data(7,:)=[3 1 1 6 1 8 1];
data(8,:)=[3 2 1 3 1 7 1];

Such a description begins our code (named **resistor.m**), as shown. Notice that the very first line of this description defines the variable *numberofnodes,* which is equal to the number of nodes *in the entire circuit.* This is distinct from the first element in each of the rows of *data,* which is the number of nodes connected to the xth node.

```
%resistor.m
for currentnode=3:numberofnodes
  nynodes=data(currentnode,1);
  cx=0;
  for loop=3:2:2*nynodes+1;
     cx=cx+1/(data(currentnode,loop));
  end
  for loop=3:2:2*nynodes+1
      pxy=1/(cx*data(currentnode,loop));
      data(currentnode,loop)=pxy;
  end
  for loop=5:2:2*nynodes+1;
      data(currentnode,loop)
        =data(currentnode,loop) . . .
        +data(currentnode,loop-2);
  end
  data(currentnode,loop)=1;
  end
  V=zeros(1,numberofnodes);
  V(1)=10000000;V(2)=0;
for startnode=3:numberofnodes
  for walk=1:10000000
      currentnode=startnode;
      while currentnode>2
          keeplooking=1;
```

```
23              decision=rand;
24              look=3;
25              while keeplooking==1
26                  if decision
                          <data(currentnode,
                          look)
27                        keeplooking=0;
28                        currentnode
                          =data(currentnode,
                          look-1);
29                  else
30                        look=look+2;
31                  end
32              end
33          end
34          if currentnode==1
35              V(startnode)=V(startnode)+1;
36          end
37      end
38  end
39  V/10000000
```

Lines 01 through 15 transform the matrix *data* into a *probability transition matrix*, using (5.5.2) and (5.5.5).[13] What that means is that the *xth* row of *data* will, when Lines 01 through 15 have finished, no longer specify the resistances through which node *x* is connected to other nodes. Instead, the elements in *data* where the resistance values *were* located will now contain numbers that determine the probabilities that a random walker at node *x* will move to each of those other (connected) nodes. Specifically, these probabilities are not the actual node transition probabilities, but rather are *cumulative* probabilities. For example, the new row in *data* for row 3 becomes

$$data(3,:)=[3\ 1\ 0.3333\ 4\ 0.6666\ 8\ 1],$$

To understand this, we next turn to the actual random walk portion of the code (Lines 18 through 38). The variable *decision* is assigned a random value from 0 to 1 (Line 23), and the code then sequentially uses *decision* to decide which node to go to next. If, for example, the walk is currently at node 3 ($currentnode = 3$) then $data(3, :)$ says to go to node 1 if $decision < 0.3333$, but if not then go to node 4 if $decision < 0.6666$, but if not then go to node 8 (notice that $decision < 1$ is certain to occur at this point because of Line 14, and that will be the case with probability 0.3334).

As mentioned earlier (note 13), Line 16 defines the row vector V of length *numberofnodes*. $V(i)$ will, when the code finishes, be the number of times (out of ten million walks) a walk starting at node i reaches node 1 before it reaches node 2. To convert the elements of V to probabilities (and, hence, to voltages), we divide by ten million. An individual random walk, beginning at *startnode* (which runs from 3 to *numberofnodes*—see Line 18), continues until Line 21 determines the walk has either reached node 1 or node 2 (all the other nodes are, of course, greater than 2). Lines 34 through 38 determine which it is and, if it is node 1, then $V(startnode)$ is incremented by 1. Ten million such walks are launched from every node from 3 to *numberofnodes*.

Okay, now to the ultimate test of our code: *does it actually work*? To answer that, let's give **resistor.m** the four circuits we analyzed earlier, one after the other, and compare what we got theoretically for the node voltages with what the code computes. So, starting with the 3D-resistor cube, with the circuit description given earlier, in just a few seconds **resistor.m** returned the node voltages

$$v_3 = 0.6000 \quad v_4 = 0.4000 \quad v_5 = 0.5997$$
$$v_6 = 0.3999 \quad v_7 = 0.6000 \quad v_8 = 0.3999,$$

in excellent agreement with (5.6.1).

Turning next to the pinwheel of Figure 5.3.1, the circuit's description is

```
numberofnodes=7;
data(3,:)=[3 1 1 2 1 4 1];
data(4,:)=[3 2 1 3 1 5 1];
```

```
data(5,:)=[3 2 1 4 1 6 1];
data(6,:)=[3 2 1 5 1 7 1];
data(7,:)=[3 2 1 6 1 1 1];
```

and again, in just seconds, **resistor.m** returned the node voltages

$$v_3=0.3889 \quad v_4=0.1667 \quad v_5=0.1109$$
$$v_6=0.1666 \quad v_7=0.3889,$$

in excellent agreement with (5.3.1), (5.3.2), and (5.3.3).

We next consider the ladder circuit of Figure 5.4.1, with the description

```
numberofnodes=8;
data(3,:)=[3 1 1 4 1 5 1];
data(4,:)=[3 2 1 3 1 6 1];
data(5,:)=[3 3 1 6 1 7 1];
data(6,:)=[3 4 1 5 1 8 1];
data(7,:)=[2 5 1 8 1 0 0];
data(8,:)=[2 6 1 7 1 0 0];
```

and in just seconds **resistor.m** gave the result

$$v_3=0.6340 \quad v_4=0.3659 \quad v_5=0.5368$$
$$v_6=0.4637 \quad v_7=0.5123 \quad v_8=0.4880,$$

values again in excellent agreement with (5.4.1) through (5.4.6).

And finally, if we present **resistor.m** with the description of the lattice circuit of Figure 5.4.3, that is, with

```
numberofnodes=8;
data(3,:)=[4 1 1 2 4 5 2 6 4];
data(4,:)=[4 1 4 2 4 5 8 6 2];
data(5,:)=[4 3 2 4 8 8 4 7 4];
data(6,:)=[4 3 4 4 2 7 16 8 1];
data(7,:)=[3 5 4 6 16 8 4 0 0];
data(8,:)=[3 5 4 6 1 7 4 0 0];
```

the code quickly returned the node voltages

$$v_3=0.7608 \quad v_4=0.5981 \quad v_5=0.7089$$
$$v_6=0.6683 \quad v_7=0.6909 \quad v_8=0.6788,$$

in outstanding agreement with (5.4.8)—*and without having to do a matrix inversion*!

5.7 Epilogue

Well, great, *the code works,* and we can use it to help in understanding the water flow question that opens this chapter. If you recall the analogy mentioned there (relating water flow in pipes to electricity flowing in a resistor network), you can now appreciate what mathematicians call the *Rayleigh monotonicity law.* Named after English mathematical physicist John William Strutt (1842–1919), who had the inherited title of Lord Rayleigh, the law states if we have *any* resistor circuit energized by a battery, then increasing any of the resistors in the circuit will always *decrease* the battery current. We can use **resistor.m** to experimentally test this claim.[14]

For example, you will recall from section 5.2 that the body-diagonal battery current in a 3D-resistor cube of one-ohm resistors is 1.2 amperes with a 1-volt battery. The code would calculate the current as (using the notation of Figure 5.6.1)

$$\frac{1-v_3}{r_{13}}+\frac{1-v_5}{r_{15}}+\frac{1-v_7}{r_{17}},$$

and so if we increase *any* resistor in the cube, the battery current should *decrease.* For example, suppose all the resistors are one ohm (in particular, we have $r_{13}=r_{15}=r_{17}=1$), but we change $r_{82}=r_{28}$ from 1 to 10. Then the circuit description becomes

```
numberofnodes=8;
data(3,:)=[3 1 1 4 1 8 1];
data(4,:)=[3 2 1 3 1 5 1];
data(5,:)=[3 1 1 4 1 6 1];
```

```
data(6,:)=[3 2 1 5 1 7 1];
data(7,:)=[3 1 1 6 1 8 1];
data(8,:)=[3 2 10 3 1 7 1];
```

and the code computes the battery current to be 0.96 amperes, which is *significantly* less than 1.2. This is of course no *proof* of the monotonicity law (far from it!), but it does offer the opportunity to have some fun playing with the circuit parameters, as well as offering me a natural place to end this chapter.

6

Simulation of a Purely Deterministic Problem

6.1 Introduction

The previous four chapters may have given you the incorrect impression that computer simulations necessarily involve issues of randomness. To show you that is not so, this chapter and the next cover two distinctly different problems that are each totally random-free yet lend themselves nicely to simulation. Indeed, both will severely challenge you if you attempt analytical solutions. Simulation is not just an alternative approach for both but, in my opinion, is the *only* approach for the problem in this chapter! Let me begin with the concept of *computation by Turing machine*.

The subject of Turing machines is generally associated with highly theoretical studies in computer science, and not so much (if at all) with the everyday concerns of "practical" computer users. It would be incorrect, however, to assume there are no problems of great interest even to those who view Turing machines as simply the playground for would-be pure mathematicians. (For a bit more on this, see the box at the end of the chapter.) The end result for this chapter will, in fact, show you a theoretical Turing machine problem that, despite being understandable by a fifth grader, would nonetheless drive (most) PhD mathematicians crazy—but which we will answer, *exactly*!, with a computer simulation that takes less than a second to run (and about five minutes to write and type).

6.2 Turing Machines

After completing his undergraduate degree at Cambridge University, English mathematician Alan Turing (1912–1954) came to America in 1936 to start his PhD studies at Princeton University. During his two years there, his work in what would today be called computer science was closely watched by von Neumann, who was so impressed with Turing he offered him a job as his assistant. Realizing that war was about to come to his homeland, the recently anointed *Doctor* Turing declined and returned to England in 1938, where he conducted his famous work in helping to defeat the fantastically complex German military Enigma code of World War II.[1]

The part of Turing's work that mostly interested von Neumann had actually been done before he arrived at Princeton; it dealt with an exploration of the theoretical limits of machine computation and, in particular, Turing's description of what has come to be called *Turing machines*. A Turing machine is a deterministic, well-defined arrangement of hardware that computes some mathematical function. Every possible function that can be computed can be associated with some specific Turing machine.

To understand the concept of a Turing machine, we first need to discuss the idea of the *state* of a fundamental electronic building block called a *flip-flop*. It is not necessary to get into the electrical details of the flip-flop (it can be constructed from relays as in the ASCC, in vacuum tubes as in MANIAC-I, or in solid-state devices like the transistors and integrated circuitry of the computers that came even later); rather, the *logical* behavior of a flip-flop (as a function of its inputs) is all a machine designer needs.

A flip-flop is a circuit that, depending on its inputs, is either ON or OFF. That is, the flip-flop is a binary circuit like a switch. It is, at every instant of time, in one of its two possible states. If you build a digital machine using *n* flip-flops, the machine has the possibility of being, at any instant of time, in one of the 2^n different ON/OFF possible states of the *n* flip-flops. MANIAC at Los Alamos and the IAS machine at Princeton each used thousands of what electronic engineers call *dual-triode vacuum tubes* (which simply means a single glass envelope contains two electrically identical but independent structures, each consisting of three electrodes).[2] A flip-flop is ON when one of those structures conducts current and the other doesn't and is OFF when the two structures flip (or flop!) roles. With one such dual tube, one could make a flip-flop.

Not all the tubes in those early machines were used to make flip-flops, but lots were. So, for example, if $n = 1,000$ then $2^{1,000}$ machine states is a pretty impressive number, strongly hinting at the potentially vast complexity of even the first primitive computers.

Since for any physically constructable machine *n* is finite, we say that anything we build with *n* flip-flops is a *finite-state machine.* Actually, the phrase (or some variation of it) that digital machine designers commonly use is that their creations are *sequentially clocked, finite-state, synchronous machines.* That's because a machine with 2^n possible states can switch from one state to another state *only* at prescribed instants of time, instants determined by an oscillator called the *clock* that generates a periodic sequence of pulses. We don't have to say anything more about such details here, and I tell you even this much simply to use terms you will encounter in other books.[3]

In 1936 Turing introduced his now famous sequential, clocked, finite-state machine. A *Turing machine* is the combination of a sequential, finite-state, synchronous machine with an external read/write memory storage system called the *tape* (think of a ribbon of magnetic tape). The tape is a linear string of squares, with each square holding one of several possible symbols. Most generally, a Turing machine can recognize any number of different symbols, but here I will assume the two-symbol binary case (0 or 1). In

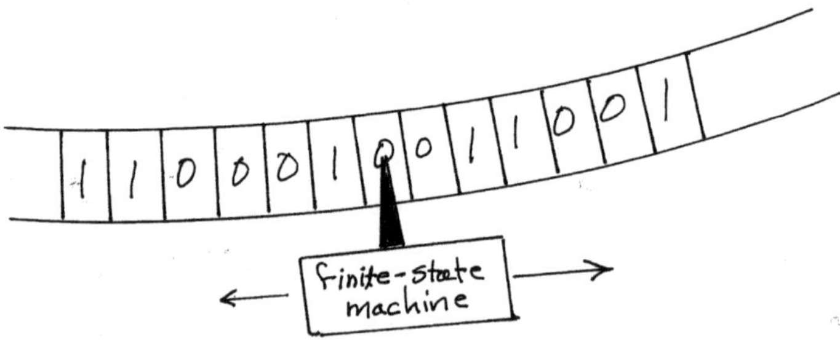

FIGURE 6.2.1. A Turing machine.

1956 Shannon showed that this assumption in no way limits the power of what a Turing machine can do.[4]

The tape is arbitrarily long in at least one or perhaps both directions. The finite-state machine is connected to a *read/write head,* which at each machine cycle (to be defined in just a moment) is located over a square on the tape. The head does three distinct operations during a machine cycle (these three operations, in fact, together with a fourth, final operation, *define* a machine cycle). First, the head reads the symbol on the square it is over; next it overwrites that symbol (perhaps with the same symbol); and then the head moves at most one square (that is, to either the left or the right neighbor square, or it *doesn't move* and so remains over the current square). Depending on both its present state and the symbol just read, the fourth and final operation of a machine cycle occurs when the finite-state machine transitions to a new state (which may, in fact, be the current state). Then, a new machine cycle begins. Figure 6.2.1 shows the connection of a finite-state machine, the read/write head (the solid triangle), and the tape. The entire arrangement, all together, is what we call a Turing machine.

When a Turing machine is placed into operation, we imagine that the tape is initially blank (that is, the symbol 0 is on all of the tape's squares)—except for a finite number (perhaps zero) of squares that have 1s. By convention, we always take the initial state as state 1, with all other states numbered upward from 2 on. And finally, we must specify over which square of the tape the read/

write head is initially placed. The finite-state machine and the read/write head then move along the tape (we imagine that the tape is motionless) according to the internal details of the finite-state machine and the particular sequence of symbols encountered on the tape.

A Turing machine's power to compute comes not from super technology but from its tape. Because of the tape's arbitrarily long length, a Turing machine has the ability to "remember" what happened in the arbitrarily distant past. The price paid for that ability to retrieve information from the unlimited past is *time*: clunking along at one tape square per machine cycle means that to return to a previously visited remote square could take a lot of time. Generally, in fact, compared to the everyday digital computers we are all used to, Turing machines are very slow. They can get the job done, yes, but you had best be prepared to wait for the answer! As a pioneer in computer science, MIT mathematician Marvin Minsky (1927–2016), so nicely put it, despite the "staggering inefficiency of a Turing machine, it is possible to execute the most elaborate possible computational procedures with Turing machines whose fixed structures [the finite-state machine and the read/write head] contain only dozens of parts. One can imagine an interstellar robot, for whom reliability is the prime consideration, performing its computations in such a leisurely manner over eons of spare time."[5]

At some (perhaps far-distant) point in time after we turn on a Turing machine, it presumably completes its task (whatever that might be), and the finite-state machine enters the special state of 0 (called the *halting state*) and the Turing machine stops.

Now, what does it mean to say a Turing machine *computes*? Here is a simple example. It seems reasonable that for the claim to hold that there is a Turing machine that can do any task a modern digital computer can do, we would demand there should certainly be a Turing machine able to *add*! In response to that expectation, Figure 6.2.2 shows what is called the *state-transition diagram* for a Turing machine that, with one exception to be mentioned soon

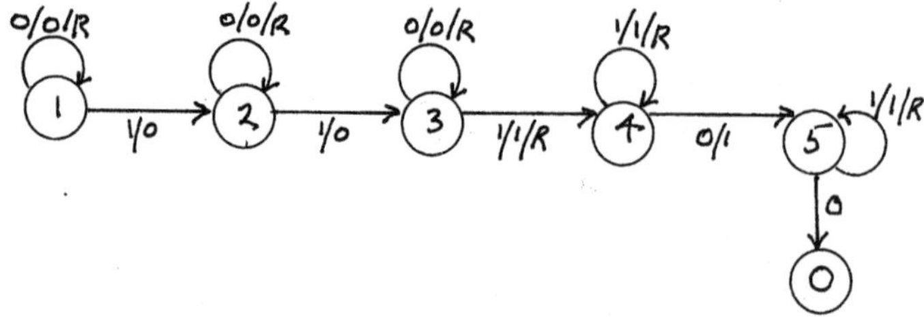

FIGURE 6.2.2. The state-transition diagram for a Turing machine adder. With six states, we could build this machine with three flip-flops.

(if you simply can't wait, look ahead to note 6), adds any two nonnegative integers m and n that are placed on the tape, leaves their sum on the tape, and then halts. I will explain the notation in the figure shortly. Our convention for representing m and n on an initially blank tape (that is, all 0s) is to write $m+1$ consecutive 1s for m, then a 0, and then $n+1$ consecutive 1s for n. All the rest of the tape is still blank. This way of coding m and n is called *unary notation*. The reason for using one additional 1 for the values of m and n is to allow for m and n to be zero (which is represented by a single 1). This Turing machine is said to *compute* the function $f(m, n) = m+n$.

The circles with numbers inside are the states, and the lines with arrowheads represent the transitions from state to state, with each such directed line marked with the conditions causing that transition. For example, 0/0/R means "read 0, write 0, move right," and 1/0 means "read 1, write 0, don't move." Notice that the read/write head in this machine never moves left. State 0 is a halting state. This machine is so simple that, starting in state 1 with the read/write head over any tape square to the left of the first 1, you should be able to follow the state-transition diagram step-by-step, with pen and paper, to confirm the following three test cases:

$0+2=$ ... 0101110 ...
which should give ... 01110 ...

2 + 0 = ... 0111010 ... which should give ... 01110 ...
2 + 3 = ... 0111011110 ... which should give ... 01111110 ...

For these small values of *m* and *n* it is not very hard to follow the operation of the machine by hand. Yet for larger values of *m* and *n* it does become a bit tedious (to say nothing of the boring task of just initializing the tape with the unary codes for large *m* and *n*). The simulation code of **add.m** does all that tedious work for us. The writing of the code is easy to do, being simply an obvious replication of the state-transition diagram. This is important to understand because we will write a similar code for the final problem of this chapter (the one I teased you about in the beginning), a problem so monstrously convoluted that to do it by hand is beyond all probability. So, here is the code and a walk-through of **add.m**.

```
%add.m
m=23;n=34;tape=zeros(1,100);state=1;
location=1;
for j=2:m+2
  tape(j)=1;
end
for j=m+4:m+4+n
    tape(j)=1;
end
while state>0
     symbol=tape(location);
     if state==1
          if symbol==0
               location=location+1;
          else
               tape(location)=0;state=2;
          end
     elseif state==2
          if symbol==0
```

```
18                location=location+1;
19            else
20                tape(location)=0;state=3;
21            end
22        elseif state==3
23            if symbol==0
24                location=location+1;
25            else
26                location=location+1;state=4;
27            end
28        elseif state==4
29            if symbol==0
30                tape(location)=1;state=5;
31            else
32                location=location+1;
33            end
34        else
35            if symbol==1
36                location=location+1;
37            else
38             state=0;
39               end
40        end
41    end
42    sum(tape)
```

Line 01 sets the values of *m* and *n*, initializes a vector named *tape* consisting of 100 elements of all zeros (the 100 is large enough to ensure the read/write head won't run off the right end of the tape), initializes the state of the finite-state machine to 1, and sets the read/write head location to being over the first square at the left end of the tape. Lines 02 through 07 do the unary coding on the tape for the values of *m* and *n*. Lines 08 through 41 are simply the state-transition diagram written in words. So, as long as the code hasn't entered state 0 (the halting state), Line 09 gets the value of the symbol on the tape square that the read/write head is over, and then (if in state 1) *if* the symbol is 0, all that happens is that the read/write head moves right. But *if* the symbol is 1, then Line 14 writes 0 at the current location, the read/write head does *not* move, and the state is set to 2. And so on for all the other states. When the code finally enters state 0, Line 42 is executed, where *sum* (the MATLAB instruction that sums the values of the elements of the *tape* vector) produces the correct answer of 58 (the code for the number of 1s in the unary code for the answer 57).[6]

6.3 The Busy Beaver Challenge

In a 1962 paper[7] that has become a minor computer science classic, Hungarian-born mathematician Tibor Radó (1895–1965) described a Turing machine problem that is not only technically important in theoretical computer science but also fun. We will concentrate here on the fun part. Radó's goal was the discovery of, for every positive integer value of *n*, the binary Turing machine with *n* states (plus a halting state), starting with a blank tape, that eventually halts (after a *finite* number of read/write head moves) with the largest possible number of 1s on the tape. The requirement to halt after a finite number of moves is to dismiss as trivial any machine that simply writes 1 on every tape square the read/write head visits as it moves endlessly along the tape. Radó called the number of 1s after the halt the *score* of the machine.

So, for example, suppose $n = 3$. For that case, Radó wrote

> Consider a potentially both-ways infinite tape where each square contains a 0. Start [the] machine described in [Figure 6.3.1] [reading] any square. The reader will find that the machine stops after a few shifts, and when it stops, there are six ones on the tape.

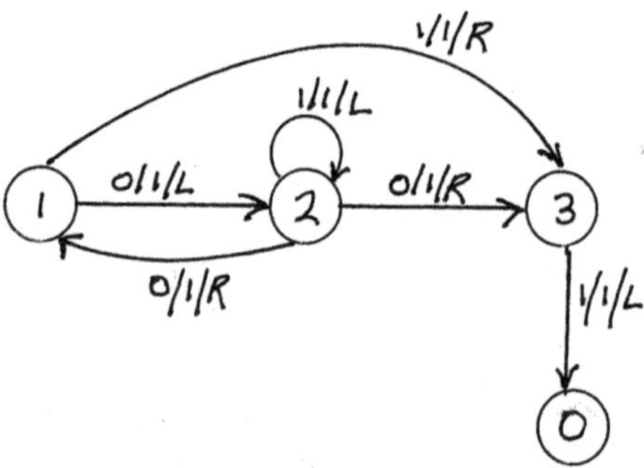

FIGURE 6.3.1. Radó's 3-state Turing machine.

When Radó wrote, he conjectured that this machine's score of 6 was the largest that any $n = 3$ state machine could achieve. That speculation was proven the following year in a doctoral dissertation in mathematics at The Ohio State University (Radó was the student's advisor), and in 1965 the two published their proof in the formal literature.[8] Radó termed his machine a *busy beaver* (who says mathematicians have no sense of humor!). In Radó's notation, now commonly used by computer scientists, $\Sigma(3) = 6$. The machine of Figure 6.3.1 is simple enough that you can, with a little patience and care, confirm its score of 6 by hand, as well as discover that Radó's "few shifts" are 21 shifts. This is usually written as $S(3) = 21$. Notice, carefully, that in Radó's machine the read/write head *always*, on *every* machine cycle, moves either left (L) or right (R). Unlike the ladder of Figure 6.2.2, where the read/write head has the option under certain circumstances of not moving, there is no such option for Radó's machine. You will soon see why Radó made this a requirement.

Of course, why bother to confirm by hand when you know (from the previous section) how easy it is to write a computer simulation. The code **BB3.m** does the job, with the additional twist to what we did earlier, with **add.m**, being the variable *shift* which, starting from an initial value of 0, is incremented after the completion of *every* machine cycle. Remember, for a busy beaver machine, the read/write head *must* move (there is no rest for a busy beaver!).

```
%BB3.m
state=1;tape=zeros(1,51);
location=25;shift=0;
while state>0
symbol=tape(location);
    if state==1
        if symbol==0
            tape(location)=1;location=
            location-1;state=2;
        else
            location=location+1;state=3;
        end
    elseif state==2
        if symbol==0
            tape(location)=1;location=
            location+1;state=1;
        else
            location=location-1;
        end
    else
        if symbol==0
            tape(location)=1;
            location=location+1;state=2;
        else
          location=location-1;state=0;
        end
    end
  shift=shift+1;
end
sum(tape),shift
```

In his 1962 paper, Radó described the general *busy beaver problem* this way, in the form of a competition (this is the fun part of the chapter):

i. The contestant selects a positive integer n; and then makes up his own n -state Turing machine.
ii. He starts his machine [in state 1] on an all-zero tape, and satisfies himself that his machine stops after a certain number of s shifts.
iii. He then submits his entry, as well as the shift-number s, to any member (in good standing) of the International Busy Beaver Club.
iv. [That member] first verifies that the entry actually stops exactly after s shifts. . . . If the entry fails to stop after s shifts, it is rejected; if it stops after fewer than s shifts, it is returned to the contestant for correction. After the entry has been verified, its score is the number of ones on the tape when it stops.

Radó then notes that "the reader . . . will soon realize the difficulties involved in this sort of problem. Beyond the enormous number of cases to [consider], he will find that it is very hard to see whether certain entries do stop at all. This is the reason for the requirement that each contestant must submit the shift number s with his entry [my emphasis]."

Here is what Radó meant by his remark on an "enormous number of cases" for an n -state binary Turing machine. The description for such a machine means that, for each nonhalting state the following must be specified at each machine cycle (which begins by reading the tape symbol at the read/write head's current location). To be specific, suppose that symbol is 0. Then the machine must decide which symbol to use to replace the 0 (either a 0 or a 1, and so there are two choices). Then the machine has to decide which way to move, and since a busy beaver *must* move, then again there are two choices. Finally, the machine must decide what its new state will be (there are $n+1$ choices, including going to the halting state). So, if the read symbol under the read/write head is 0, there are

$$(2)\,(2)\,(n+1)=4(n+1)$$

possibilities for each of the n nonhalting states (the halting state has no such choices to make). That is, there is a total of $[4(n+1)]^n$ possibilities. This same reasoning holds if the read symbol is 1. Thus, the total number of possibilities is $[4(n+1)]^n\,[4(n+1)]^n=[4(n+1)]^{2n}$.

This expression grows *very* fast as n increases. For $n=3$, for example, it is 16^6 or almost 17 million, which helps explain why Lin and Radó had to turn to computers to help them search through that mountain of candidate machines. It also explains Radó's remark at a symposium held the same year the proof for the values of $\Sigma(3)$ and $S(3)$ was published: "As regards [the values of] $\Sigma(4)$ and $S(4)$, the situation seems to be entirely hopeless at present."

And yet, just three years later, computer scientist Allen Brady at the University of Notre Dame, using massive computer support, was able to *conjecture* the almost correct values (he had the correct $\Sigma(4)=13$, but his $S(4)=106$ was off by one).[9] And that is where matters stood until, seventeen years later, Brady *proved*[10] the correct values are $\Sigma(4)=13$, $S(4)=107$. The state-transition diagram for a 4-state busy beaver is in Figure 6.3.2, and its simulation code is **BB4.m**.

```
%BB4.m
state=1;tape=zeros(1,1001);
location=501;shift=0;
while state>0
symbol-tape(location);
     if state==1
          if symbol==0
               tape(location)=1;location=
               location+1;state=2;
          else
```

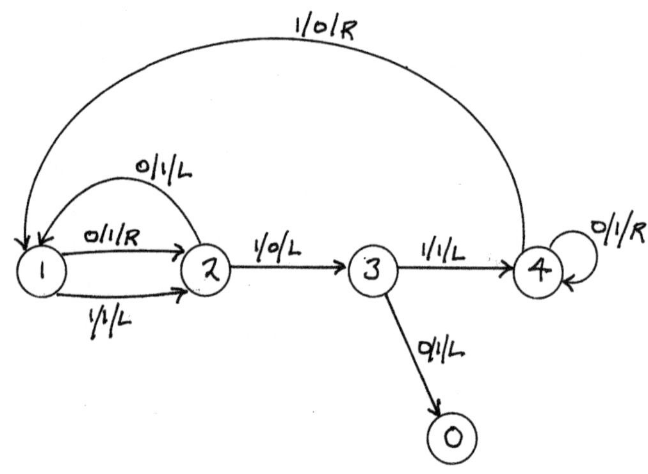

FIGURE 6.3.2. 4-state busy beaver.

```
            location=location-1;state=2;
        end
    elseif state==2
        if symbol==0
            tape(location)=1;
            location=location-1;state=1;
        else
            tape(location)=0;
            location=location-1;state=3;
        end
    elseif state==3
        if symbol==0
            tape(location)=1;
            location=location-1;state=0;
        else
            location=location-1;state=4;
        end
    else
        if symbol==0
            tape(location)=1;
            location=location+1;
```

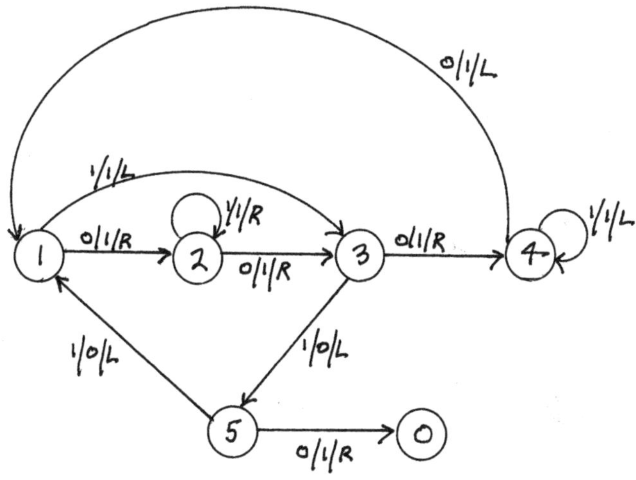

FIGURE 6.3.3. 5-state busy beaver (maybe).

```
            else
                  tape(location)=0;
                  location=location+1;state=1;
            end
      end
   shift=shift+1;
end
sum(tape),shift
```

With Brady's solution to the $n=4$ case, attention of course shifted to the $n=5$ one. There are now 24^{10} candidate machines to consider, a number so large you can best appreciate it when written out in words: *more than sixty-three trillion*. As I write (mid-2024), the best (that is, the high scorer) machine is one found decades ago (1989) by two German computer scientists.[11] Figure 6.3.3 shows its state-transition diagram. While it is conceivable that a human could run through a hand analysis for the $n=3$ and $n=4$ cases, it is impossible to imagine a human doing that for the $n=5$ case and getting the same result twice in a row. It *is* easy to imagine, however, that a human attempting that task would go crazy! The code **BB5.m** performs (in less than a second) a simulation of Figure 6.3.3, with the result being $\Sigma(5) \geq 4{,}098$ and $S(5) \geq 47{,}176{,}870$ (the reason for the $\geq$ signs is that there may be $n = 5$ machines that achieve higher values—or maybe not).

```
%BB5.m
state=1;tape=zeros(1,100000);
location=50000;shift=0;
while state>0
    symbol=tape(location);
    if state==1
        if symbol==0
            tape(location)=1;
            location=location+1;state=2;
        else
            location=location-1;state=3;
        end
    elseif state==2
        if symbol==0
            tape(location)=1;
            location=location+1;state=3;
        else
            location=location+1;
        end
    elseif state==3
        if symbol==0
            tape(location)=1;
            location=location+1;state=4;
        else
            tape(location)=0;
            location=location-1;state=5;
        end
    elseif state==4
        if symbol==0
            tape(location)=1;
            location=location-1;state=1;
```

```
            else
                location=location-1;
            end
        else
            if symbol==0
                tape(location)=1;
                location=location+1;state=0;
            else
                tape(location)=0;
                location=location-1;state=1;
            end
        end
      shift=shift+1;
    end
    sum(tape),shift
```

I will end this chapter on a whimsical note. It is almost certain that anybody whose technical interests are such that they are reading this book, has read the short fantasy tale "The Devil and Simon Flagg" by American writer Arthur Porges (1915–2006). Originally published in the August 1954 issue of *The Magazine of Fantasy & Science Fiction*, and reprinted numerous times since[12], it has achieved status as a minor classic of mathematical fiction. (Porges had the "street creds" to write such a story: a master's degree in mathematics and many years on the math faculty at Los Angeles City College.) The story tells how a professor of mathematics challenges the Devil (after some haggling over the potential loss of his soul, and of the possibility of suffering eternal damnation—you know, the usual issues that come up when dealing with the Devil[13]) to resolve Fermat's Last Theorem, and the ending is such that (perhaps surprisingly) one can't help but feel a bit of sympathy for the Devil. Still, the story *has* lost some of its original punch because the solution to Fermat's Last Theorem was finally accomplished decades later. All that is needed, however, to restore the story to its full glory is a new (unsolved) math problem. Some have suggested proving the Riemann Hypothesis (see my book *In Pursuit of Zeta-3*, Princeton 2023, for what that is all about), but I can think of an even better choice. I suggest the challenge to the Devil should be to calculate the values of $\Sigma(100)$ and $S(100)$. The task of searching through $404^{200} > 10^{521}$ binary Turing machines (a number that *stupendously dwarfs* the number of subatomic particles in the visible universe) should keep even His Satanic Majesty "busy as a beaver" (and *perhaps* out of trouble) until hell freezes over.

One of the theoretical problems traditionally associated with Turing's work is the famous *halting problem*. That curious name refers to the question concerning the possibility of writing a computer program that, when presented with *any* program (along with that program's input), can decide if the submitted program (and its input) will, once started, eventually halt or will, instead, run "forever.' The answer is *NO*, such a wonderful program is *impossible* to write. Like a unicorn, you can imagine its existence but it is still just a fantasy. And that is too bad because, if it did exist, we could use it to check our programs to see if they contain any dreaded *infinite loops*, the result (one not at all uncommon, sad to say) of faulty programming that results in a computer getting stuck in an endless loop. You can find a discussion (my elaboration of the discussion in Minsky's book, note 5) of the halting problem, along with an elegant proof—due *not* to Turing, but to American mathematician Martin Davis (1928–2023)—in my book *Number-Crunching: Taming Unruly Computational Problems in Mathematical Physics and Science Fiction*, Princeton 2011, pp. 330–334.

A beautiful exposition, at the level of this book, on the connection of Turing machines and the halting problem, was written by Davis as one of the essays ("What Is Computation?") in the book *Mathematics Today—Twelve Informal Essays*, Springer 1978, pp. 241–261. Good reading, too, is the paper by Salvador Lucas, "The Origins of the Halting Problem," *Journal of Logical and Algebraic Methods in Programming*, June 2021.

7

Schwartz's Question

THE TOUGHEST PROBLEM IN THIS BOOK

7.1 Introduction

The purpose of the title of this chapter is to intentionally tickle your curiosity, so let me tell you a little of the history of what we will be doing here. Way back one morning in 2010, I opened that day's *Boston Globe* newspaper and, in particular, its insert of (the now defunct) *Parade Magazine*. A regular feature of *Parade* at that time was a column by well-known puzzle writer Marilyn vos Savant. There I read the following question, posed by one of her readers (identified only as "M. Schwartz," in Ventura, California):

> A friend and I once went from his house to mine with one bike. I started walking and he took the bike. When he got a couple of blocks ahead, he left the bike on the sidewalk and started walking. When I got to the bike, I started riding, passing him, and then left the bike a couple of blocks ahead. When he got to the bike, he started riding. We did this the whole way. At least one of us was always walking. At times one was riding; at other times we were both walking. I'm sure this was faster than if we had no bike. But some people insist that it was no faster because somebody was always walking. Who was right?

Marilyn's answer was brief, to the point, and correct.[1] "The reader is right. It's true that someone was always walking, but neither friend walked the whole distance. Both biked part of the way. This increased their average speed, and so they saved time." On the surface, this is a very simple problem, but it contains a deeper one (*much* deeper, in fact, than I originally appreciated), one that Marilyn either didn't spot or, if she did, thought it too complicated for her readers and so decided not to pursue it. As I finished reading her answer I remember thinking, "This could be the basis for a really neat problem for the new book, where *new book* referred to my then-in-progress *Number-Crunching* (published the next year by Princeton).

That proved to be the case, and here is what I eventually wrote in that book's Chapter 7 (titled "The Leapfrog Problem"):

> What's so special about "a couple of blocks ahead"? Would some other separation distance be "better"? It seems physically obvious that the "best" (I'll sharpen what *that* means soon) separation between the friends (the distance before the bike-riding boy gets off and starts walking) will be a function of all the parameters of the problem: the distance between the two houses (d), the walking speed (w) and the riding speed (r).[2] I also think it reasonable to assume that $r > w$. . . . Let's call the individual trip times for the boys t_1 and t_2; it should be clear that the two boys will, in general, *not* finish the trip simultaneously. That is, in general, $t_1 \neq t_2$. If we call the longer of these two times the *duration* of the trip, the criterion for what *best* means is the value of the separation distance that gives the smallest value for the duration. That is, if we denote the separation distance by h, then the optimal h is the value of h that *minimizes the maximum* of t_1 and t_2.

To gain an appreciation for the timing of a leapfrog trip, suppose each boy walks at 2 mph and rides at 6 mph, and that the distance to be traveled is one mile. To walk the whole way would take $\frac{1}{2}$ hour or 1,800 seconds, while to ride the whole way would take $\frac{1}{6}$ hour or 600 seconds. So, using the bike, the boys would like to find the separation distance they should use to achieve the smallest trip duration time (which we now know will be some value between 600 seconds and 1,800 seconds).

At this point I naively thought I could simply write down a few equations involving all the variables and use those equations to express h as some function that could be minimized by taking a derivative and setting it equal to zero (the typical "freshman calculus" approach to an extreme problem). To my growing concern, however, I found myself unable to do that. I was stunned. The problem is completely transparent *physically* and yet, try as I did, I couldn't even get started *mathematically*. Frustrated, I at last calmed down and decided to do what this book's theme preaches: even though an analytical approach was apparently going to require more insight and/or firepower than I possessed, the very transparency of the physics would allow me to *simulate* the travels of the two boys. (That experience may, in fact, have planted the seed in my mind that eventually grew into this book.) And that is what I did, arriving at the code **leapfrog.m** in *Number-Crunching*.

That original computer code simulated, a large number of times, the physics of the boys leapfrogging their trip, each time using a different separation distance: the duration of each trip (along with the separation distance used) was saved. A plot of duration versus separation distance (for the values of the walking and riding speeds in the previous paragraph) produced Figure 7.1.1. To say I was greatly surprised by that plot is a huge understatement—I recall thinking, "Can this be right?" Now almost two decades later, I don't recall *what* I expected to see for a plot of duration versus separation distance, but the rising, sawtooth curve of Figure 7.1.1 was surely not it.

At the risk of sounding pompous, my emotional reaction to Figure 7.1.1 was almost like the reaction the great American physicist Richard Feynman (1918–1988) reported when he made a

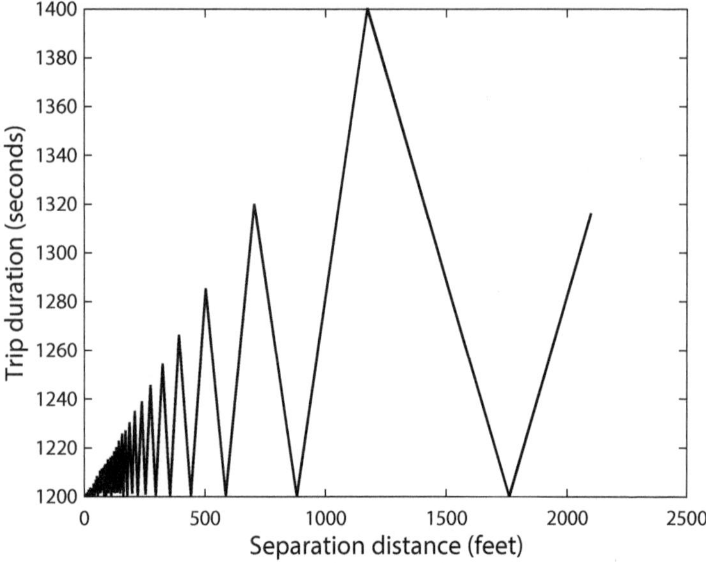

FIGURE 7.1.1. Trip duration versus separation distance for a one-mile trip, with each boy walking at 2 mph and riding at 6 mph.

significant breakthrough concerning a physical puzzle that was then driving everybody in the physics world nuts: "I was very excited. It was the first time, and the only time . . . that I knew [something] that nobody else knew."[3] This is not to compare the leapfrog problem to the deep physics problem that Feynman studied, but rather to say the emotional euphoria I had was the same as Feynman's.

The plot in Figure 7.1.1 was *not* created by the original leapfrog computer code in *Number-Crunching*, but rather by the new code to be developed later in this chapter. This new code allows the walking and riding speeds of the two boys to be different (see note 2), but when run with equal walking times and equal riding times the result is identical to the *Number-Crunching* code's plot, which was a great relief! (see Figure 7.1.2). The range of h, the separation distance, varies from 1 foot (mathematically possible but admit-

FIGURE 7.1.2. The author's reaction to this book's new computer code successfully creating Figure 7.1.1, in agreement with the plot created by the **leapfrog.m** code in *Number-Crunching*. Image by permission of Clipart Of LLC (artist Ron Leishman).

tedly pretty unrealistic in practice, requiring the boys to continually be hopping on and off the bike) to 2,100 feet (about the length of seven football fields) in one-foot steps. Figure 7.1.1 tells us (perhaps surprisingly) that there are multiple values of h that result in a minimum trip duration of 1,200 seconds (the three largest values less than 2,100 feet are 587 feet, 880 feet, and 1,760 feet).

This is where I ended matters in *Number-Crunching*, leaving readers with two challenges: (1) Develop an analytical solution to the problem (see the following box) or, failing that, (2) Develop a computer simulation code in which the walking and riding speeds are independently specified (keeping the caveat of note 2 in mind). As the years passed, I received numerous e-mails from readers all around the world expressing various clever suggestions

Thinking that an analytical solution to the leapfrog problem should be "straightforward" is seductive, as the following story illustrates. After *Number-Crunching* appeared in 2011, I received an invitation to deliver that year's annual Richard W. Sampson Lecture in Mathematics at Bates College. For my final talk, an evening lecture on the value of computers in mathematics, I used the leapfrog problem as an example. As it happened, in the audience was a professor of mathematics at nearby Colby College (Bates and Colby are premiere private liberal arts schools in Maine). After my talk the Bates math faculty treated me to an elegant dinner and, as it happened, the Colby professor was seated next to me. As we chatted, he casually commented that he was pretty sure he knew how to solve the problem analytically, but it was all just a bit too much to get into specifics over dinner. He would instead, he promised, write his solution in detail and send it to me in the next few days. A week or so later, I *did* receive an e-mail from him—with the admission he was stuck, too! He promised to keep working on it, however, as he was sure he was very close to overcoming the difficulty that was stopping him, and he would keep me informed. Alas, I never heard from him again. (I can't help but think of the poor Devil in the tale I told you about at the end of the last chapter.)

and observations about the problem. But there were no direct responses to my challenges—until mid-2019 when I received *two* nearly simultaneous analytical solutions. My two correspondents (both PhD academic mathematicians) were Dirk Nerinckx at the Catholic University in Leuven, Belgium, and Phillip Schmidt, professor emeritus of mathematics at both the University of Akron and Northern Kentucky University. You can read through their analyses on your own,[4] as we concentrate here on challenge 2, the development of a simulation code.

7.2 The New Code

To simulate the motions of the two boys and of their bike, it is vital to have a crystal-clear understanding of the various possibilities for the instantaneous locations[5] and speeds of each of the three.

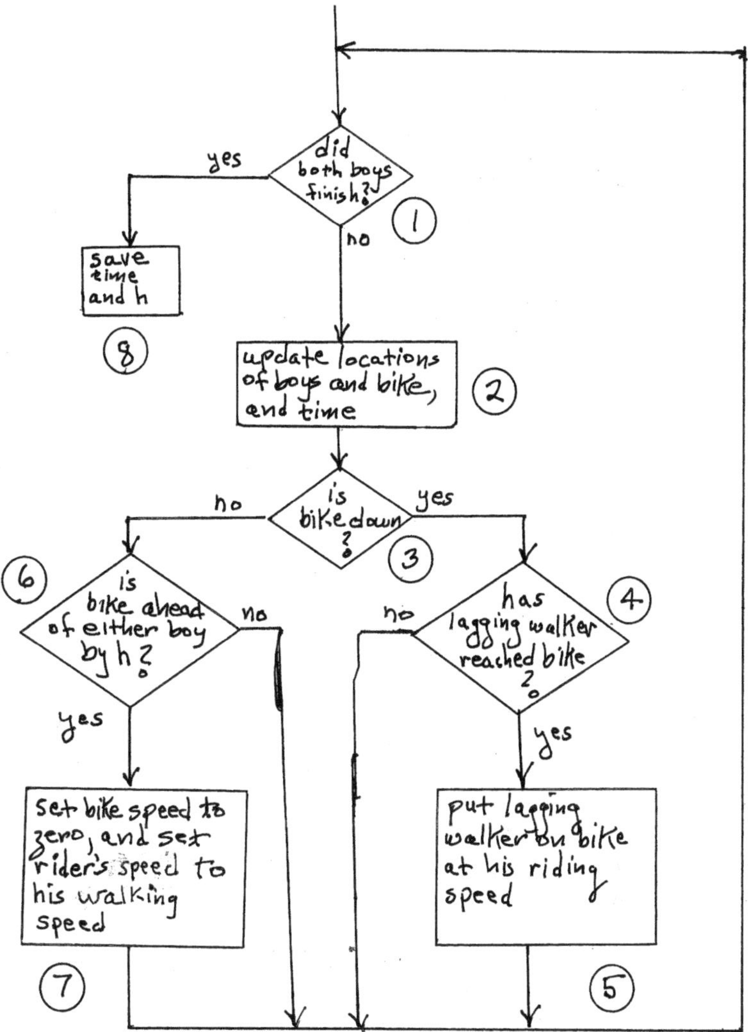

FIGURE 7.2.1. The logical flowchart for leapfrog simulation, for a given value of the separation distance, *h*.

The boys are each always moving at one of two speeds, while the bike has three possible speeds (the riding speed of the one currently on the bike, or zero when both boys are walking). To keep all that straight, I found it extremely helpful *not* to immediately jump into writing code, but rather to start by creating the logical flowchart of Figure 7.2.1.[6] The circled numbers next to each box in that figure are keyed to specific lines of the new code (to be called **LF.m**).

To understand the code, you need to be familiar with the following: if we label the two boys as *boy*1 and *boy*2, then we define the elements of the four vectors *x*, *ws*, *rs*, and *sep* as follows:

$$\begin{aligned} x(1) &= \text{distance traveled by } boy1 \\ x(2) &= \text{distance traveled by } boy2 \\ x(3) &= \text{distance traveled by the bike} \\ ws(1) &= \text{walking speed of } boy1 \\ ws(2) &= \text{walking speed of } boy2 \\ rs(1) &= \text{riding speed of } boy1 \\ rs(2) &= \text{riding speed of } boy2 \end{aligned}$$

$$sep(1)=1, sep(2)=2, sep(3)=3, \ldots, sep(2100)=2100.$$

We next define the variable *lastonbike* as follows:

$$lastonbike = \begin{cases} 1 \textit{ if } boy1 \textit{ was the last to ride the bike} \\ 2 \textit{ if } boy2 \textit{ was the last to ride the bike} \end{cases}.$$

Finally, we define the following three additional variables:

$$\begin{aligned} b1s &= \text{current speed of } boy1 \\ b2s &= \text{current speed of } boy2 \\ bikes &= \text{current speed of the bike} \end{aligned}$$

Our last assumption is that when a simulation starts, using a given value for the separation distance h (the distance the bike rider is *ahead* of the walker that triggers the rider getting off the bike), the code will take *boy*1 as the initial bike rider and *boy*2 as the initial walker. To reverse that order, we simply reverse the entries in the *ws* and *rs* vectors. Depending on the walking and riding speeds of the boys, the decision of who rides first *can* make a difference, as you will see when we run the code. The code **LF.m** is given in the following box, and you have seen the MATLAB commands in it before, with one exception. In Line 06 the command is of the form *while expression1|expression2,* where the vertical bar denotes a logical inclusive-OR test. That is, the *while* loop of Lines 06 to 32 will be executed *only if at least one* of the two expressions is satisfied (this implements the decision in ① of Figure 7.2.1). In other words, the *while* loop will *not* be executed if *both* expressions *fail* to be satisfied (which occurs when *both* boys have finished the trip).

```
%LF.m
01  d=1;sep=linspace(1,2100,2100);
    ws=[2 2];rs=[6 6];dt=0.001;
02  d=d*5280;ws=ws*5280/3600;
    rs=rs*5280/3600;
03  for loop=1:2100
04        h=sep(loop);x=[0 0 0];t=0;
05        lastonbike=1;b1s=rs(1);b2s=ws(2);
          bikes=b1s;
06        while x(1)<d|x(2)<d
07              x(1)=x(1)+b1s*dt;
08              x(2)=x(2)+b2s*dt;
09              x(3)=x(3)+bikes*dt;
10              t=t+dt;
11              if bikes==0
12                       if min(x(1),x(2))>=x(3)
13                          if lastonbike==1
14                                b2s=rs(2);
15                                lastonbike=2;
16                          else
17                                b1s=rs(1);
18                                lastonbike=1;
19                          end
20                          bikes=rs(lastonbike);
21                       end
22              else
23                    if x(3)-min(x(1),x(2))>=h
```

```
24                      bikes=0;
25                      if lastonbike==1
26                          b1s=ws(1);
27                      else
28                          b2s=ws(2);
29                      end
30                  end
31              end
32          end
33          time(loop)=t;
34      end
35      plot(sep,time,'k-')
```

Line 01 sets the values of all the parameters that, once set, never change from one simulated trip to the next simulated trip. Those parameters include *d* (the length of the trip, in miles); the vector *sep* that holds 2,100 values of *h*, in feet, that will be used in the 2,100 trip simulations; the walking and riding speed vectors, *ws* and *rs*, in mph; and *dt* (the time increment). Line 02 converts *d* to feet, and the walking and riding speed vectors to feet/second. The *for* loop of Lines 03 to 34 runs the code through 2,100 trip simulations. For each such simulation, Lines 04 and 05 set the initial values of all the parameters that *do* change for each individual simulation: the separation distance *h*, the locations of the boys and the bike (the elements of the vector *x*), which boy was the last one on the bike (*lastonbike*), and the current speeds of the boys and the bike, *b1s*, *b2s*, and *bikes*).

As mentioned earlier, Line 06 checks to see if *both* boys have traveled distance *d*; if so, the leapfrogging is over, the *while* loop is skipped, and ⑧ of Figure 7.2.1 is implemented in Line 33. Lines 07 through 10 execute ②. Line 11 is the decision of ③ that sends the code either to ④ or to ⑥. If to ④, Lines 12 through 15 implement ⑤, or if to ⑥, Lines 23 through 31 implement ⑦.

Well, okay, we now have a code. But is it *correct*? Up until I received the e-mails from Nerinckx and Schmidt (note 4), I had no way to check a simulation result against a theoretical analysis.

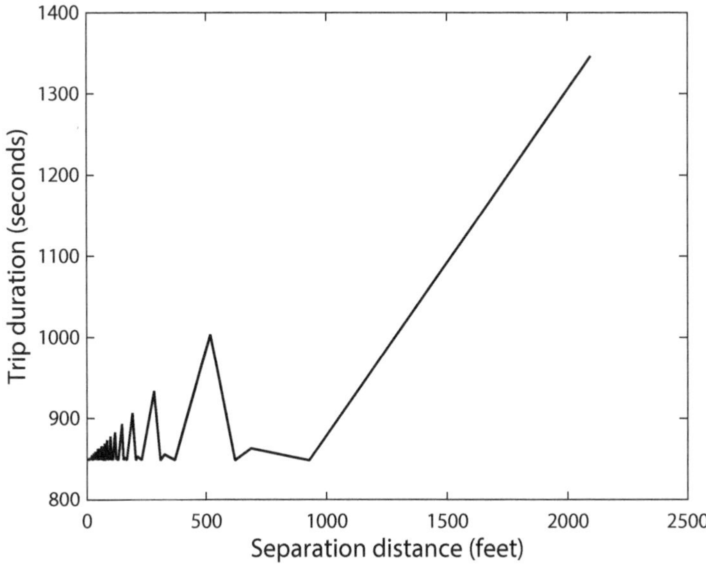

FIGURE 7.2.2. *ws*[4 2] and *rs* = [5 8].

With Schmidt's analysis (in particular) in hand, however, I at last had a theoretical result against which to check **LF.m**. In his paper, Schmidt plotted his theoretical result for the specific case of *boy1* and *boy2* walking at speeds 2 mph and 4 mph, respectively, and riding at speeds 8 mph and 5 mph, respectively. Assuming *boy1* is the initial rider (as is assumed in **LF.m**), we would then write *ws* = [2 4] and *rs* = [8 5]. In his theoretical analysis, however, Schmidt assumed that *boy2* is the initial rider. So, to run a simulation with the **LF.m** code of Schmidt's example, we must reverse the *ws* and *rs* vectors and use *ws* = [4 2] and *rs* = [5 8]. With that alteration, **LF.m** produces Figure 7.2.2, which—long drum roll—*exactly matches* Schmidt's theoretical plot (see his Figure 4).

We can use **LF.m** to explore the question that was briefly raised earlier. What if, as Schmidt assumed, the "other boy" is the initial rider? Does that make a difference? To answer that, let's use the vectors reversed from those that produced Figure 7.2.2: *ws* = [2 4] and *rs* = [8 5]. The result is shown in Figure 7.2.3 and we see that

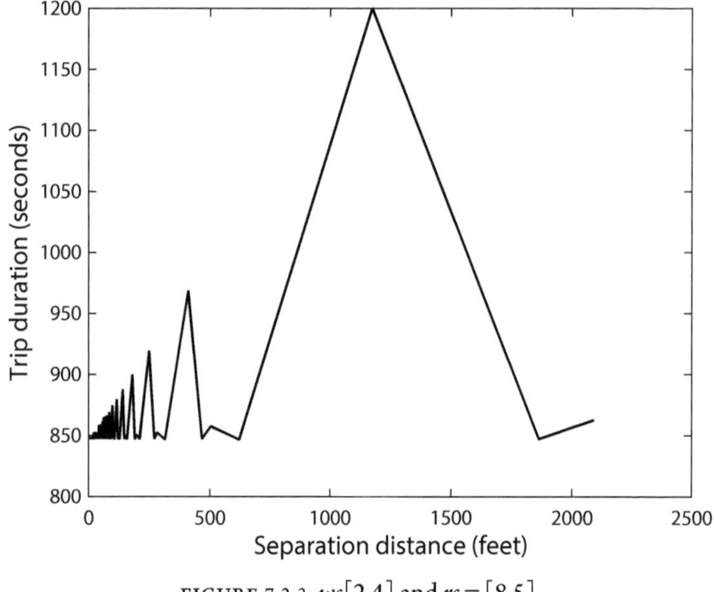

FIGURE 7.2.3. $ws[2\ 4]$ and $rs=[8\ 5]$.

the answer is both *no* and *yes*. It's *no* because there is no difference in the minimum duration (either choice for the initial rider gives a minimum duration of 847 seconds), but it's also *yes* because the critical values for h that result in that minimum duration *do* depend on who rides first.

Professor Schmidt ends his paper with a challenge of his own. What if, he asks, the two boys use different values for the separation distance that triggers each to get off the bike? In particular, suppose the boys agree that when each sees his friend in the distance get off the bike, it will then be a walk for each that requires the same time to reach the bike. That is, a "slow" walker will have a shorter distance to walk to reach the bike, and a "fast" walker will have a longer distance to walk. If they agree that each wants the walk to the dropped bike to take T seconds, then the separation distance used by *boy*1 is $ws(2)T$ and the separation distance used by *boy*2 is $ws(1)T$.

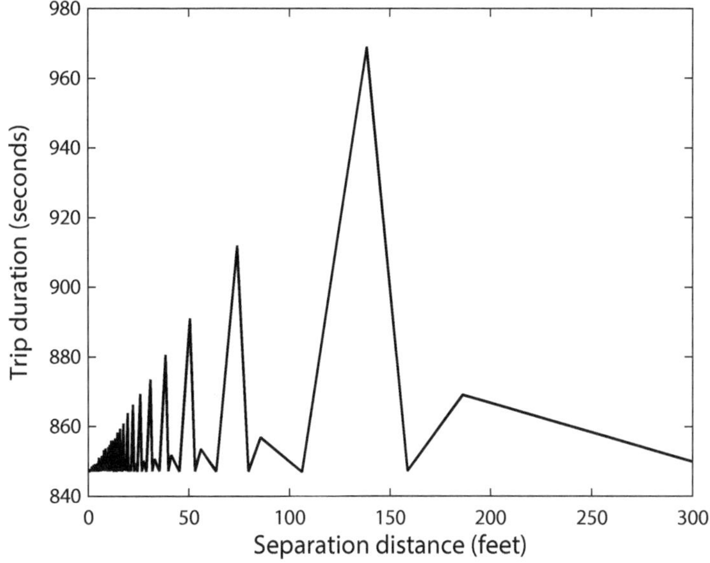

FIGURE 7.2.4. *ws* = [4 2] and *rs* = [5 8].

So, there is Schmidt's challenge: analyze this modified leapfrogging method to plot the duration of the trip as *T* varies up to a maximum (let's say) of five minutes (300 seconds). Professor Schmidt doesn't provide a theoretical solution, but to modify **LF.m** isn't a difficult task. I won't list my code here, but Figure 7.2.4 shows the result for *ws* = [4 2] and *rs* = [5 8]. See if you can confirm that figure. Notice that the minimum duration is still 847 seconds.

8

A Counterexample Concerning Computer Simulations

A drunk man will *always* find his way home. A drunk bird, on the other hand, might not find its way back to its nest.

—A LITTLE JOKE TOLD BY PROBABILITY ENTHUSIASTS ABOUT RANDOM WALKS (THIS WILL MAKE A *LOT* MORE SENSE BY THE END OF THIS CHAPTER!)

8.1 Introduction

The previous two chapters may have given you the impression that *any* well-defined physical process, random or deterministic, can be studied by simulation. This chapter will show you, in the spirit of honesty (and contrary to the main thrust of this book), that such a conclusion may be a bit too hasty! Here I will discuss a physical process which is easy to set up as a computer simulation yet is unable to answer an immediately obvious question about that process. We *can* get an answer, but it will require a theoretical treatment. The intent of this chapter is to illustrate that while computer simulation is a powerful tool that often can take you around mathematical difficulties to arrive at an answer, you had best pay attention to sharpening your mathematical skills, too!

So, here is the problem. In 1921 George Pólya (introduced in note 3 in the Prelude) studied what are called *random walks on rectangular lattices*.[1] In one dimension, the walk can be thought of as a man randomly moving from integer to integer along a line. In two dimensions, the walk can be thought of as the man moving randomly up and down and back-and-forth among the points in the plane with integer coordinates. In three dimensions, think of a child on a jungle gym. Here, we will limit ourselves to the one- and two-dimensional cases.

Starting at $x=0$ at time $t=0$, the walker takes one step to the right with probability p, or one step to the left with probability q, where $p+q=1$. At $t=1$ the walker does the same thing, and so on for $t=2$, $t=3$, etc. Now, suppose we write p and q as $p=\frac{1}{2}+\epsilon$ and $q=\frac{1}{2}-\varepsilon$, where $0\leq\epsilon\leq\frac{1}{2}$. The $\varepsilon=0$ case (with $p=q=\frac{1}{2}$) is called the *symmetrical* walk, while $\varepsilon\neq 0$ describes a walk with *bias* (or *drift*).[2] This is a very well-defined physical process, and to write a computer code that, starting with the variable $x=0$, increasers x by 1 with probability $\frac{1}{2}+\epsilon$ or decreases x by 1 with probability $\frac{1}{2}-\epsilon$ (for any given legal value of ε) is easy to do (as in the following code named **feller.m**).

```
%feller.m
eps=0;prob=0.5+eps;x=0;
horz=linspace(1,100000,100000);vert=horz;
for loop=1:100000
     if prob>rand
          x=x+1;
     else
          x=x-1;
     end
     vert(loop)=x;
end
plot(horz,vert,'-k')
```

In a famous book on probability theory published in the late 1960s, Croatian-American mathematician William Feller

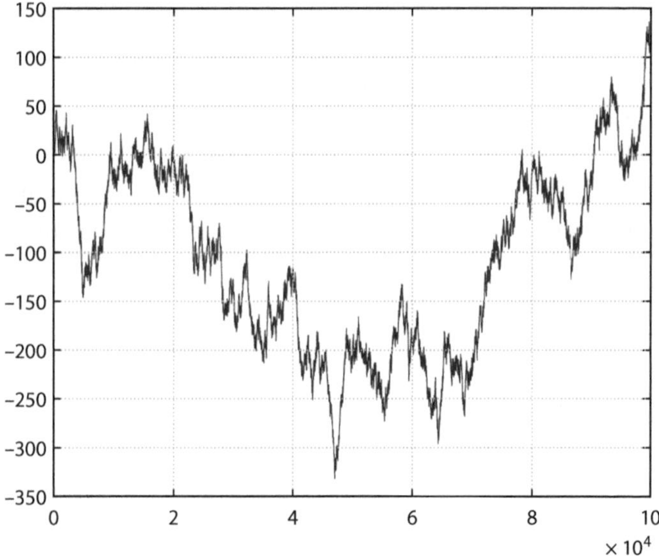

FIGURE 8.1.1. Feller's experiment for a fair coin ($\varepsilon = 0$) flipped 100,000 times (the vertical axis shows the imbalance in the number of times one side shows compared to the other side).

(1906–1970) interpreted the $\varepsilon = 0$ case as the flipping of a *fair* coin, with the value of x representing the number of times one side of the coin has shown compared to the other side. Feller showed he was ahead of many of his theoretical colleagues by including a *computer experiment* of 10,000 flips that displayed a startling property of the symmetrical walk. In the code **feller.m** I have repeated Feller's experiment (with a tenfold increase in the number of flips to 100,000), with the result shown in Figure 8.1.1. There we clearly see that, even with a *fair* coin (ε is *eps* in **feller.m** set equal to zero), one side of the coin shows *significantly more often* than does the other side. To most people, this appears to violate our concept of what *fair* means. As Feller wrote, "This *looks* absurd [but if this result seems] startling, this is due to our faulty intuition and to our having been exposed to too many vague references to a mysterious 'law of averages.'"[3] There is, in fact, no such law.[4]

If you look at Figure 8.1.1 for a while you will notice that even though the value of x at times deviates from zero, it eventually returns to $x=0$ and then deviates again only to again return after a while to $x=0$. Every time you run **feller.m** you will of course get a plot different in detail (this is a *random* process, after all!), but this same general behavior is always observed. This prompts the following question: If the random walker starts at $x=0$ at time $t=0$, is it possible that after the walker has moved left-and-right along the x -axis (possibly returning one or more times to the starting point) that eventually there comes a time after which she *never again* returns to $x=0$?

A computer simulation can't answer that question because it would require the simulation to run *forever*! If it has been a trillion steps since the last return to the origin, that in no way forbids the possibility that if you let the simulation run for another trillion steps the walk might again return to $x=0$. No matter how long it has been since the last return, maybe if you just ran the simulation for a few more years you would see another return. Or maybe not. A theoretical analysis is the only way to definitively answer our question, and the answer is truly astounding.

As Pólya showed in his 1921 analysis, the answer is NO for $\varepsilon=0$ (a one-dimensional *symmetrical* walk will *eternally* return to $x=0$). But an $\varepsilon\neq 0$ one-dimensional walk *can* (as mathematicians put it) *escape to infinity*.[5] The answer to our question switches from NO to YES *discontinuously* as ε goes from zero to nonzero: such discontinuous behavior simply can't be captured by a computer code that runs for a finite time.[6]

8.2 Two Mathematical Preliminaries

This section is based on a discussion in an elegant little monograph by Doyle and Snell (see note 10 in chapter 5, and in particular Doyle and Snell's pages 119–121). I have elaborated a bit, here and there, on that discussion because Doyle and Snell wrote for a readership just a bit more advanced than I am assuming for this book. As you read this section, notice, *carefully*, that there is no assumption about the dimensionality of the walk anywhere in the analysis. Our results here will, in fact, hold for a one-dimensional walk, a two-dimensional walk, and a three-dimensional walk; indeed, they will hold for a walk in a space of *any* dimension.

Let u denote the probability the random walk, starting at the origin, eventually returns to the origin. Thus, $1-u$ is the probability the walk *never* returns to the origin. So, the probability the walk is at the origin exactly k times (including the time the walk starts) is given by $u^{k-1}(1-u)$. If we write r as the expected number of times the walk is at the origin (see the Appendix for more on the expected value of a random variable) then

$$r=\sum_{k=1}^{\infty}ku^{k-1}(1-u)=(1-u)\sum_{k=1}^{\infty}ku^{k-1}. \qquad (8.2.1)$$

To evaluate the sum at the far right of (8.2.1), write

$$\begin{aligned} S=\sum_{k=1}^{\infty}u^k &= u+u^2+u^3+\cdots \\ &= u(1+u+u^2+\cdots) \end{aligned} \qquad (8.2.2)$$

and notice that the sum in (8.2.1) is the derivative of S:

$$\sum_{k=1}^{\infty}ku^{k-1}=\frac{dS}{du}. \qquad (8.2.3)$$

Now

$$1+u+u^2+\cdots=\frac{1}{1-u},\quad |u|\le 1,$$

as can be seen by simply cross multiplying (because u is a probability we know $0\le u\le 1$, which satisfies the inequality condition on u). Thus, from (8.2.2) we have

$$S=\frac{u}{1-u} \qquad (8.2.4)$$

and so (8.2.3) states

$$\frac{dS}{du} = \frac{(1-u)-u(-1)}{(1-u)^2} = \frac{1}{(1-u)^2} = \sum_{k=1}^{\infty} k u^{k-1}.$$

Combining this result with (8.2.1) we arrive at

$$r = (1-u)\frac{1}{(1-u)^2} = \frac{1}{1-u}. \tag{8.2.5}$$

This simple result (our first mathematical preliminary) has a profound interpretation: if $u = 1$ then $r = \infty$. That is, if the probability of the walk returning to the origin is one, then the expected number of returns is infinitely large. That is just another way of saying the walk returns to the origin *eternally*. If you wait long enough since the last time the walk was at the origin, you *will* see it there again (and again, and again, and . . .).

Now, let's look at the walk in a different way. The symbol u was used earlier (in deriving our first result) to denote the probability the walk eventually returns to the origin. Let's slightly modify our notation to write u_k to denote the probability the walk returns to the origin *in exactly k steps*, where $k \geq 1$. Further, let's define a sequence of random variables, R_k, where $R_k = 1$ if the walk is at the origin after k steps, and $R_k = 0$ if the walk is *not* at the origin after k steps. We can then write the total number of times the walk returns to the origin (the value of which is the random variable we will call R) as the sum of the individual R_k. That is,

$$\boldsymbol{R} = \boldsymbol{R}_1 + \boldsymbol{R}_2 + \cdots = \sum_{k=1}^{\infty} \boldsymbol{R}_k \tag{8.2.6}$$

and so the expected number of times the walk returns to the origin—what we wrote in (8.2.1) as r—is given by

$$r = E(\boldsymbol{R}) = E\left\{\sum_{k=1}^{\infty} \boldsymbol{R}_k\right\} = \sum_{k=1}^{\infty} E(\boldsymbol{R}_k). \tag{8.2.7}$$

The expected value of each of the R_k is easy to calculate:

$$E(R_k) = (1)u_k + (0)(1-u_k) = u_k.$$

Thus, (8.2.7) becomes

$$r=\sum_{k=1}^{\infty}u_k. \tag{8.2.8}$$

With (8.2.8) we now have a new result (our second mathematical preliminary), along with (8.2.5). We saw from (8.2.5) that $r=\infty$ means the walk returns eternally to the origin, and so (8.2.8) now gives us the mathematical signature for *any* walk that eternally returns: a walk (*any* walk, in a space of *any* dimension) returns to the origin eternally if

$$\sum_{k=1}^{\infty}u_k=\infty. \tag{8.2.9}$$

If (8.2.9) doesn't hold, then the walk will not exhibit eternal return and there will come a time after which you will never see the walk at the origin again. In the next section we will calculate what the u_k actually are for a one-dimensional walk (for any value of ε).

8.3 The One-Dimensional Walk

For a one-dimensional walk to be back at the origin after k steps, it is physically obvious that the walk has taken as many steps to the right as it has to the left. That is, k must be even; if k is odd, $u_k=0$. So, we are interested in calculating u_{2n}, for $n\geq 1$, the probability the walk has returned to the origin in exactly $2n$ steps. The $2n$ steps can occur in any order, with each possible order having probability p^nq^n. There are $\binom{2n}{n}$ different ways to have n steps to the right and n steps to the left, and therefore

$$u_{2n}=\binom{2n}{n}p^nq^n=\frac{(2n)!}{(n!)^2}p^nq^n. \tag{8.3.1}$$

So, there is our problem, the calculation of

$$\begin{aligned}\sum_{n=1}^{\infty}u_{2n}&=\sum_{n=1}^{\infty}\frac{(2n)!}{(n!)^2}\left(\frac{1}{2}+\varepsilon\right)^n\left(\frac{1}{2}-\varepsilon\right)^n\\&=\sum_{n=1}^{\infty}\frac{(2n)!}{(n!)^2}\left(\frac{1}{4}-\varepsilon^2\right)^n.\end{aligned} \tag{8.3.2}$$

Applying Stirling's asymptotic formula for factorials—see (1.3.2) in chapter 1—we have

$$\frac{(2n)!}{(n!)^2}=\frac{\sqrt{2\pi 2n}(2n)^{2n}e^{-2n}}{(\sqrt{2\pi n}n^n e^{-n})^2}=\frac{\sqrt{4\pi n}2^{2n}n^{2n}e^{-2n}}{2\pi n n^{2n}e^{-2n}}$$

$$=\frac{2\sqrt{\pi}\sqrt{n}2^{2n}}{2\pi n}=\frac{1}{\sqrt{\pi n}}2^{2n}$$

and so (8.3.2) becomes

$$\sum_{n=1}^{\infty}u_{2n}=\sum_{n=1}^{\infty}\frac{1}{\sqrt{\pi n}}2^{2n}\left(\frac{1}{4}-\varepsilon^2\right)^n$$

$$=\sum_{n=1}^{\infty}\frac{1}{\sqrt{\pi n}}2^{2n}\left[\left(\frac{1}{4}\right)(1-4\varepsilon^2)\right]^n$$

$$=\sum_{n=1}^{\infty}\frac{1}{\sqrt{\pi n}}2^{2n}\left[\left(\frac{1}{2^2}\right)(1-4\varepsilon^2)\right]^n$$

$$=\sum_{n=1}^{\infty}\frac{1}{\sqrt{\pi n}}2^{2n}\frac{1}{2^{2n}}[(1-4\varepsilon^2)]^n$$

or,

$$\sum_{n=1}^{\infty}u_{2n}=\frac{1}{\sqrt{\pi}}\sum_{n=1}^{\infty}\frac{(1-4\varepsilon^2)^n}{\sqrt{n}}. \tag{8.3.3}$$

The question of returning to the origin for a one-dimensional walk thus reduces to determining the value of (8.3.3). If (8.3.3) is infinite, the walk exhibits "eternal return," while if (8.3.3) is finite, the walk can escape to infinity. For $\varepsilon=0$ (the symmetrical walk) the answer is easy: the sum diverges and so the one-dimensional symmetrical walk eternally returns to the origin. This is so because $\sqrt{n}<n$ for all $n>1$ and therefore

$$\sum_{n=1}^{\infty}\frac{1}{\sqrt{n}}>\sum_{n=1}^{\infty}\frac{1}{n}=\infty$$

as the sum on the right of the inequality is the famous harmonic series, long known (for *centuries*) to diverge.

Next, what if $\varepsilon > 0$? The case of $\varepsilon = \frac{1}{2}$ (the largest possible value for ε) is mathematically trivial, as that means every term in (8.3.3) is zero, and so the sum is zero. It is physically trivial, too, as $\varepsilon = \frac{1}{2}$ means $p = 1$ and $q = 0$, and so *every* step of the walk is in the same direction, which insures no return to the origin. Of far greater interest is the case of $0 < \varepsilon < \frac{1}{2}$.

For $0 < \varepsilon < \frac{1}{2}$ every term in (8.3.3) goes to zero not only because of the increasing denominator, but also because $1 - 4\varepsilon^2 < 1$, and so the numerator of each term *decreases* with increasing n. The terms in the sum of (8.3.3) go to zero faster than they do for the $\varepsilon = 0$ case (where we found the sum diverged), but is that increased rate of decrease enough to make the sum in (8.3.3) converge? The answer is *yes,* and an easy way to prove that is to use the *integral test* (taught in freshman calculus). If we write $c = 1 - 4\varepsilon^2$ (and so $0 < c < 1$), then our question is

$$\sum_{n=1}^{\infty} \frac{c^n}{\sqrt{n}} = ? \quad \text{for} \quad 0 < c < 1.$$

The integral test states the divergence/convergence of the sum parallels the divergence/convergence of the integral

$$\int_1^{\infty} \frac{c^x}{\sqrt{x}} dx.$$

Since $\sqrt{x} > 1$ for all $x > 1$, we can write

$$\int_1^{\infty} \frac{c^x}{\sqrt{x}} dx < \int_1^{\infty} c^x dx = \int_1^{\infty} e^{ln(c^x)} dx = \int_1^{\infty} e^{xln(c)} dx$$
$$= \frac{e^{xln(c)}}{\ln(c)} \Big|_1^{\infty} = -\frac{e^{ln(c)}}{\ln(c)} = -\frac{c}{\ln(c)}$$

which is a positive (using the fact that $ln(c) < 0$ because $c < 1$) finite value. So, (8.3.3) is finite for all $\varepsilon > 0$ and the one-dimensional walk does *not* exhibit eternal return. (Read the last paragraph of section 8.1 again.) Thus, if we pick $\varepsilon > 0$, no matter by how little the value differs from zero, the nature of the one-dimensional walk

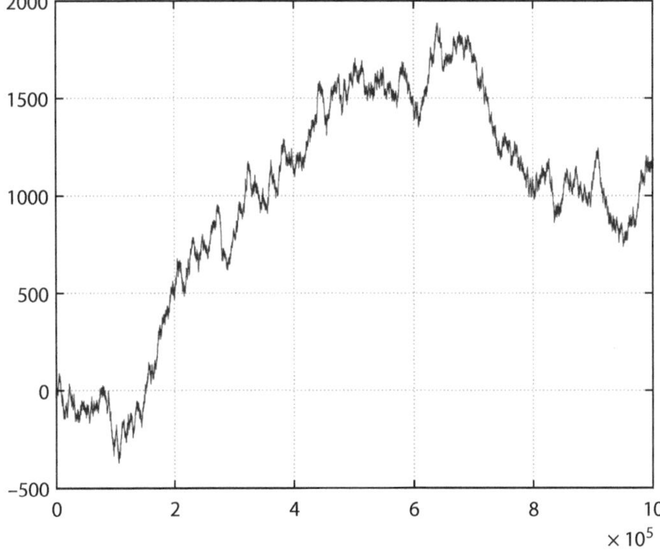

FIGURE 8.3.1. Feller's experiment for $\varepsilon = 0.0001$ (one million flips).

discontinuously changes from exhibiting eternal return to that of escaping to infinity.

We can use **feller.m** to see if the code at least hints at this curious behavior. Running the code with *eps* set to 0.0001 (and so $p = 0.5001$ and $q = 0.4999$), we get Figure 8.3.1, which is *markedly* different in nature from Figure 8.1.1 (for $\varepsilon = 0$). This is not a proof of the discontinuous behavior—only our theoretical analysis can make that claim—but nevertheless it is still an impressive example of the value of a computer in the study of a mathematical problem.

9

Monte Carlo Integration

[T]he Monte Carlo method is the real workhorse of present-day high-dimensional integration.

—FRANCIS Y. KUO AND IAN H. SLOAN, "LIFTING THE CURSE OF DIMENSIONALITY"[1]

9.1 Introduction

One of the first electronic computer applications of the Monte Carlo idea was the numerical evaluation of definite integrals. The standard numerical method (taught for centuries to beginning calculus students worldwide) is Simpson's rule that, in its most elementary form for a one-dimensional integral, approximates the area under the integrand of the integral as the sum of the areas of "many" narrow rectangles. That is,

$$\begin{aligned} I &= \int_a^b f(x)dx \approx \lim_{n\to\infty} \sum_{k=1}^{n} f(a+k-x)-x, \\ x &= \frac{b-a}{n}. \end{aligned} \tag{9.1.1}$$

As long as $f(x)$ is, as mathematicians say, "reasonably well-behaved" over the interval of integration, this generally gives good results.

Simpson's rule can be extended, in an obvious way, to the cases of double, triple, and even higher-dimensional integrals. Yet, while technically correct, we quickly encounter a serious practical complication. If each dimension is divided into n intervals, the number of evaluations of the integrand in the approximation sum of (9.1.1) grows as n^d, where d is the dimension of the integral. So, for example, if $n = 100$ for doing a one-dimensional integral like (9.1.1), then we would need $100^2 = 10{,}000$ integrand evaluations for a double integration, $100^3 = 1{,}000{,}000$ integrand evaluations for a triple integration, and so on. This rapid growth has been called the "curse of dimensionality." The Monte Carlo technique avoids this curse. (Do high-dimensional integrals actually occur, or is that simply an abstract, academic concern? The paper in note 1 briefly discusses a problem that occurred in Wall Street financial circles involving an integral with the impressive dimension of 360, the number of monthly repayments on a 30-year loan with a varying interest rate.)

9.2 The Integration Algorithm and Code

Picture the integral of (9.1.1) as shown in Figure 9.2.1, where it is imagined that the entire variation of $f(x)$ over the interval $a \leq x \leq b$ has been contained within the dashed-outlined box. In the notation of the figure, $P \geq \max f(x)$ and $N \leq \min f(x)$.

As you will recall from AP- or freshman calculus, when $f(x) > 0$ the area between $f(x)$ and the x -axis is called *positive* area, while if $f(x) < 0$ the area between $f(x)$ and the x -axis is called *negative* area. So, suppose we randomly generate the coordinates of a "lot" of points *uniformly* over the interior of the dashed-outlined box (which has area $(P-N)(b-a)$). That is, suppose we generate a large number of *pairs* $x = a + (b-a) * rand, y = N + (P-N) * rand$. Each such pair can be thought of as the location of a dart tossed randomly at the box, with every dart landing somewhere *in the box*: no darts miss the box. So, suppose we let *dart* be a variable (initially set equal to zero) that denotes how many darts have so

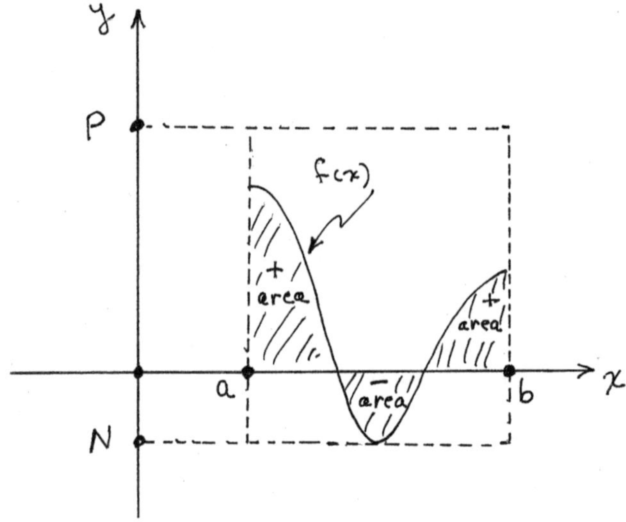

FIGURE 9.2.1. Putting the integrand in a rectangular box.

far landed between $f(x)$ and the x -axis, where we increase *dart* by 1 every time a dart lands in positive area, and decrease *dart* by 1 every time a dart lands in negative area. If we toss a total of *TD* darts at the box, then is it intuitively plausible that the ratio of *dart*/*TD* is an approximation to the ratio *I*/*area of the box*? We will assume we can write *I* as

$$I \approx (P-N)(b-a)\frac{dart}{TD}. \quad (9.2.1)$$

This *estimate* gets better[2] with increasing values of *TD*.

The code **mci.m** is (I believe) a self-evident implementation of the integration algorithm described in the previous paragraph. The code "tosses" ten million darts. Perhaps the only special note necessary is that the ampersand symbol & denotes MATLAB's logical-AND operation (in Lines 07 and 12). Line 06 defines the integrand we are integrating. For our first example, let's consider the integrand $f(x) = e^{-0.5x}\sin(3x)$, which is plotted in Figure 9.2.2,

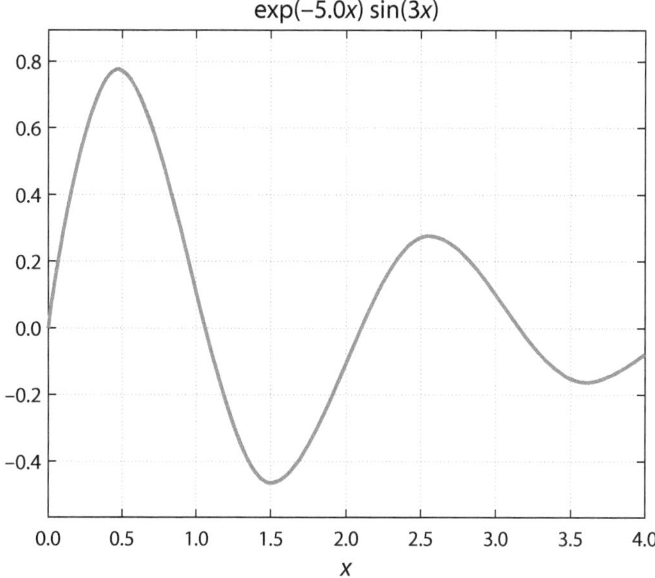

FIGURE 9.2.2. The integrand of (9.2.2).

over the interval $0 \leq x \leq 4$. From the plot we see that $a=0$, $b=4$, $P=0.8$, and $N=-0.5$. This integrand results in a doable integral (do it by parts, or just look it up in math tables). So we will have a theoretical result to compare with what the code states. That is, consider

$$\int_0^4 e^{-0.5x} \sin(3x)dx = 0.2912.... \tag{9.2.2}$$

Running the code numerous times gave results varying from 0.2903 to 0.2919, an error of less than ±0.3%.

```
%mci.m
TD=10000000;dart=0;
a=0;b=4;P=0.8;N=-0.5;
for loop=1:TD
        x=a+(b-a)*rand;
```

```
        y=N+(P-N)*rand;
        f=exp(-0.5*x)*sin(3*x);
        if f>0&y>0
            if f>y
                dart=dart+1;
            end
        else
            if f<0&y<0
                if y>f
                    dart=dart-1;
                end
            end
        end
    end
    (P-N)*(b-a)*dart/TD
```

9.3 An "Advanced" Integral

Now, to really put **mci.m** to a test, let's do an "advanced" integral. To be specific, on page 127 of Ruel V. Churchill's classic 1960 book *Complex Variables and Applications,* he assigns, as a challenge problem, the derivation of the formula

$$\int_0^{\pi} e^{\cos(x)} \cos(\sin(x))dx = \pi. \tag{9.3.1}$$

Establishing (9.3.1) is a task that is far and away beyond anything taught in freshman calculus! Indeed, Professor Churchill's suggestion is to use the powerful technique of contour integration in the complex plane, an area of mathematics generally considered advanced. But *you,* right now, don't have to know *anything* about contour integration in order to numerically evaluate (9.3.1) using **mci.m.**[3]

We start by examining the behavior of the integrand of (9.3.1) over the interval of integration, with Figure 9.3.1 showing the result. That figure suggests the assignments of the values $a=0$, $b=\pi$, $P=3$, and $N=0$. The results (from numerous executions of the

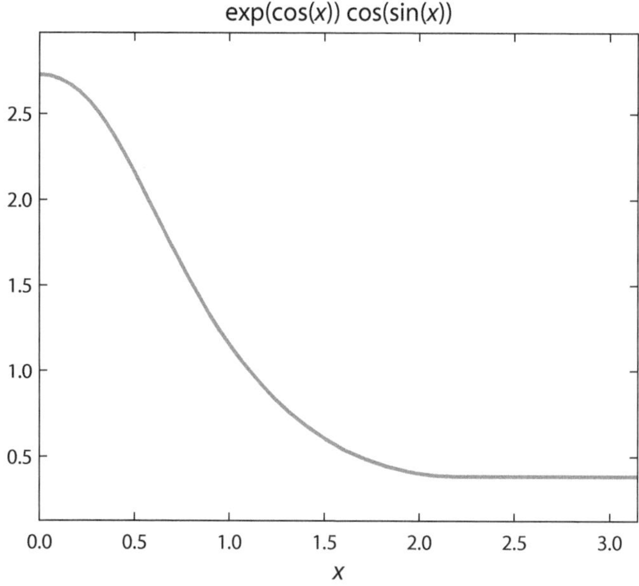

FIGURE 9.3.1. The integrand of (9.3.1).

code) gave values of the integral varying from 3.138859 to 3.145379, an error of about ±0.1%.

9.4 A Final Example

The Monte Carlo results for *one*-dimensional integrals like (9.2.2) and (9.3.1) are really no better than what we would get using Simpson's rule. It is with *higher*-dimensional integrals (as the opening quotation asserts) that Monte Carlo shines. That's because with Simpson's rule you *have* to do n^d integrand evaluations for a d-dimensional integral, while for Monte Carlo the only requirement on the number of evaluations is "the more the better." With Simpson's rule you don't get *anything* until you do *all* n^d integrand evaluations, while with Monte Carlo you get, *starting right from the get-go*, even with a small number of evaluations, ever-better results as the number of evaluations increases.

As a simple example of that wonderful feature of Monte Carlo, consider the two-dimensional (double) integral of (9.4.1):

$$\int_0^2\int_0^1(xy+3)dxdy. \tag{9.4.1}$$

This evaluates analytically as

$$\int_0^2\int_0^1(xy+3)dxdy=\int_0^2\left\{\int_0^1(xy+3)dx\right\}dy=\int_0^2\left\{\frac{1}{2}x^2y+3x\Big|_0^1\,dy\right.$$

$$=\int_0^2\left(\frac{1}{2}y+3\right)dy=\left(\frac{1}{4}y^2+3y\right)\Big|_0^2=1+6=7.$$

To do (9.4.1) with Monte Carlo, we first develop a physical interpretation of a double integral.

The integral of (9.4.1) is a particular example of the more general

$$\int_c^d\int_a^b f(x,y)dxdy. \tag{9.4.2}$$

We can visualize (9.4.2) in Figure 9.4.1, where the surface $z=f(x,y)$ "floats above" the rectangular region $a<x<b$, $c<y<d$ in the x, y -plane. (To keep things simple, I am assuming $f(x,y)\geq 0$ for all x and y. The integrand in (9.4.1), for example, satisfies that assumption.) The double-differential $dxdy$ in (9.4.2) is a tiny (differential) *area patch* in the x, y -plane that defines the cross-sectional area of a "thin" rectangular volume of height $z=f(x,y)$. That is, $f(x,y)dxdy$ is the differential *volume* of that thin volume, and the *entire integral* is the *entire volume* of the object shown in Figure 9.4.1.

So, with P denoting the maximum of $f(x,y)$, we imagine embedding the integral volume in the rectangular volume of $(b-a)$ $(d-c)P$. We then generate *triplets* of random numbers $a<x<b$, $c<y<d$, $0<z<P$ to be the location of darts in space and check to see if each triplet is above or below $f(x,y)$. That is, for each triplet we check to see if $z>f(x,y)$ or if $z<f(x,y)$. As in **mci.m**, we start with $dart=0$ and increment $dart$ by 1 if $z<f(x,y)$. If we toss a total of TD darts, the value of the integral will then be

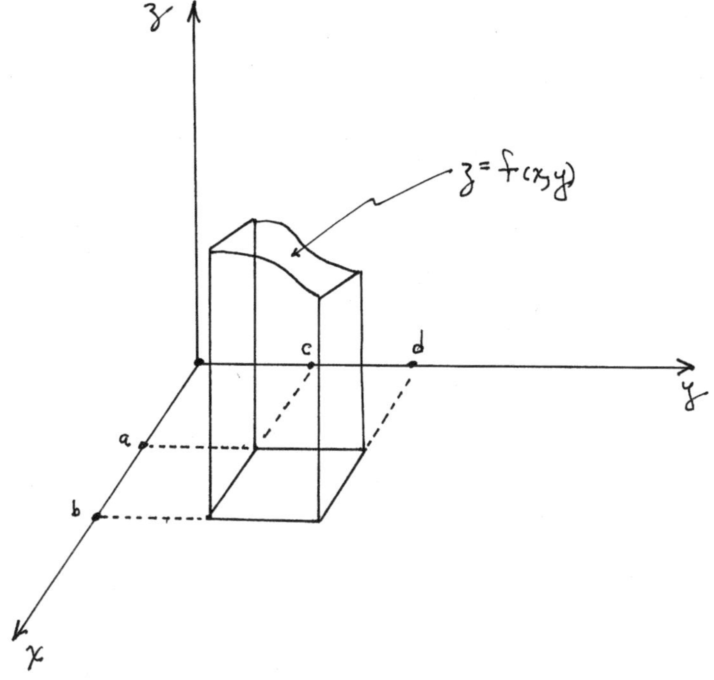

FIGURE 9.4.1. Physical interpretation of (9.4.2).

$(b-a)(d-c)P\frac{dart}{TD}$. For integrands that can be negative, then more generally we would extend our imagery to include *negative* volumes and decrease *dart* by 1 for every triplet such that $z > f(x, y)$.

For the particularly simple integral of (9.4.1) we see, by inspection, that $P = 5$ (where $x = 1$ and $y = 2$). The short, transparent code **dmci.m** (for *double Monte Carlo integration*) then performs the dart tossing of the Monte Carlo technique, with typical results shown in Table 9.4.1 for various values of *TD*.

```
%dmci.m
TD=10;dart=0;
a=0;b=1;c=0;d=2;P=5;
for loop=1:TD
```

TABLE 9.4.1. Monte Carlo Does a Double Integral

TD	Monte Carlo estimate
10	8
100	6.8
1,000	7.09
10,000	7.021
100,000	7.0086
1,000,000	7.00011

```
x=a+(b-a)*rand;
     y=c+(d-c)*rand;
     z=P*rand;
     f=x^y+3;
     if z<f
          dart=dart+1;
     end
end
(b-a)*(d-c)*P*dart/TD
```

9.5 Epilogue

To end on an amusing note, when I first wrote **dmci.m** to evaluate (9.4.1), the code did *not* converge toward 7. Rather, it converged to a value *close* to 7 (around 7.1), but nonetheless to a value sufficiently different from 7 that I couldn't wave the discrepancy away by simply mumbling "statistical sampling error." What was going on?

It was only when I sat down with a printout in large font (I have 84-year-old eyes) and looked at the code *carefully* that I realized I had not typed $f = x * y + 3$, but rather $f = x \wedge y + 3$ (the ^ symbol is MATLAB's exponentiation operator). How that happened, I have no idea—the ^ and the * are on different keys that are not even adjacent! What **dmci.m** was integrating was *not* (9.4.1), but rather

$$\int_0^2 \int_0^1 (x^y + 3) dx dy. \qquad (9.5.1)$$

This double integral is, like (9.4.1), doable:

$$\int_0^2\int_0^1 (x^y+3)dxdy = \int_0^2 \left\{ \frac{x^{y+1}}{y+1} + 3x \Big|_0^1 dy = \int_0^2 \left(\frac{1}{y+1} + 3 \right) dy \right.$$

$$= \int_0^2 \frac{dy}{y+1} + 3\int_0^2 dy = 6 + \int_1^3 \frac{du}{u} = 6 + \ln(3)$$

$$= 7.0986\ldots$$

which is pretty nearly what **dmci.m** was telling me (with just 10,000 darts, the code gave me values ranging from 7.049 to 7.176).

So, yes, it was a goof, but it was a *successful* goof!

10

Gamma-Ray Paths across a Semicircle

> A nuclear reactor is usually thought of as a region of high flux of neutrons. However, since almost all nuclear processes generating or absorbing neutrons and some involving the scattering of neutrons generate gamma-rays, there is also present a very high flux of gamma-rays. *Their absorption can cause important effects* ... [my emphasis].
>
> —W. PRIMAK, JANUARY 1956[1]

10.1 Introduction

Exposure to gamma-ray radiation is extremely dangerous to humans. Think of dental X-rays on steroids, cubed. The amount of radiation damage to the cells of the human body is related to the length of the path the radiation takes in traversing the tissue of whatever part of the body is exposed, and so the determination of path length through various tissue geometries is of great interest. One way to make such a determination is to imagine a very large number of gamma-rays randomly entering target tissue of some assumed geometry, and then calculating the path length in the tissue of each ray. A histogram of all those random path

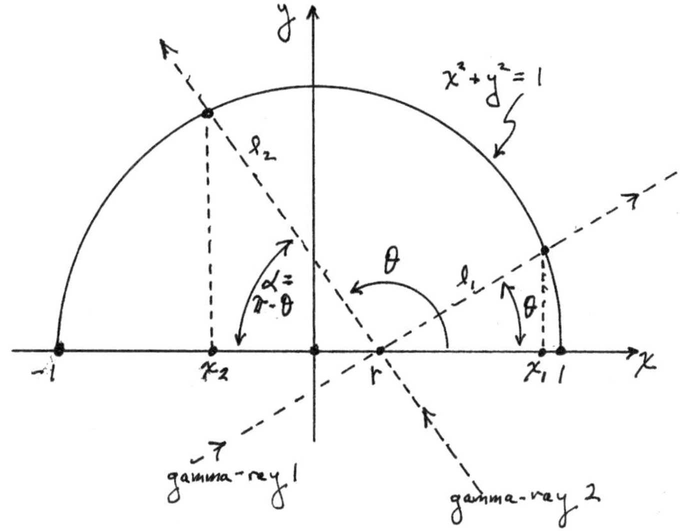

FIGURE 10.1.1. Path geometries for gamma-rays l_1 and l_1 assuming $r > 0$, for $\theta < \frac{\pi}{2}$ and $\theta > \frac{\pi}{2}$, respectively.

lengths will display the likelihood of "long" paths (those paths of highest damage).

The distribution of path lengths across a given geometry is a problem that occurs in physical situations that are, at first glance, very different from our problem here about gamma-rays. Two examples of that are (1) the path lengths of air targets being tracked by radar as they move through a surveillance space and (2) the path lengths of sound waves in a loudspeaker enclosure. The distribution in (1) is related to the probability the radar will, in fact, detect the presence of the air target, and the distribution in (2) is related to the sound quality generated by the speaker. Even though these two situations are quite different, the *mathematics* of both is identical to that of the gamma-ray problem.

The simplest geometry is probably that of a semicircle, as shown in Figure 10.1.1, where we imagine gamma-rays enter at random points along the horizontal diameter, at random angles.

In particular, if we assume the semicircle has unit radius (in whatever units you like), centered on the origin, then the random entry point of a gamma-ray is anywhere along the diameter in the interval -1 to 1. The entry angle θ has a random value in the interval 0 to 180° (π radians). In the figure, the entry point is shown at distance $r>0$ (to the right of the origin). I will say more soon about the $r<0$ case, as well as explain the angle α in Figure 10.1.1.

For any entry angle $\theta<\frac{\pi}{2}$, the gamma-ray path (the dashed line l_1) intersects the circular boundary of the tissue at a point with x-coordinate x_1 that falls to the right of the entry point. For any entry angle $\theta>\frac{\pi}{2}$, the gamma-ray path (the dashed line l_2) intersects the circular boundary of the tissue at a point with x-coordinate x_2 that falls to the left of the entry point. It appears, then, that we have *two* geometry problems here: the calculation of l_1 for $\theta<\frac{\pi}{2}$, and of l_2 for $\theta>\frac{\pi}{2}$. As you will soon see, however, the two situations are *not* as different as they may initially appear. So, let's start with the calculation of the path lengths l_1 and l_2 for the $r>0$ case.

10.2 Calculating the Path Lengths

From Figure 10.1.1 we have the slope of l_1 as $\tan(\theta)$ and so, since l_1 passes through the point $(r,0)$, we immediately have the equation of l_1 as

$$y_{gamma}=(x-r)\tan(\theta). \tag{10.2.1}$$

We can find the x-coordinate of the intersection point of l_1 with the circular boundary by setting $y_{gamma}=y_{circle}$, to get

$$(x-r)\tan(\theta)=\sqrt{1-x^2}. \tag{10.2.2}$$

The solution of this quadratic equation in x will give us the value of x_1 (see Figure 10.1.1 again).[2] If you do the algebra correctly, you should arrive at

$$x_1=r\,sin^2(\theta)\pm\cos(\theta)\sqrt{1-r^2sin^2(\theta)}. \tag{10.2.3}$$

From Figure 10.1.1 it is seen that

$$l_1 = \frac{x_1 - r}{\cos(\theta)} \tag{10.2.4}$$

which reduces, after just a few easy lines of algebra, to

$$l_1 = -r\cos(\theta) \pm \sqrt{1 - r^2 sin^2(\theta)}.$$

So, which sign do we use? Notice that for $\theta = 0$ we have the gamma-ray shooting right along the lower horizontal edge of the semicircle, entering the semicircle at $x = r$ and leaving the semicircle at $x = 1$, for a path length of $1 - r$. That immediately tells us to use the + sign,[3] and so we have our first result:

$$l_1 = -r\cos(\theta) + \sqrt{1 - r^2 sin^2(\theta)}, \quad r > 0, \quad \theta < \frac{\pi}{2}. \tag{10.2.5}$$

Next, turn your attention to the $\theta > \frac{\pi}{2}$ case and let's calculate l_2. As you can see from Figure 10.1.1, if we define $\alpha = \pi - \theta$ (and so $\alpha < \frac{\pi}{2}$), the slope of l_2 is $-tan(\alpha)$. And, since l_2 passes through $(r, 0)$, we immediately have the equation of l_2 as

$$y = (r - x) \tan(\alpha). \tag{10.2.6}$$

Notice that except for a sign, (10.2.6) looks just like (10.2.1).

Now, just as we did for l_1, we can find the x-coordinate of the intersection point of l_2 with the semicircle by setting $y_{gamma} = y_{circle}$ to get

$$(r - x)\tan(\alpha) = \sqrt{1 - x^2} \tag{10.2.7}$$

and so. . . . The dots are a nice way of saying *blah blah blah* (code for "just as we did for l_1") and we arrive at

$$l_2 = \sqrt{1 - r^2 sin^2(\alpha)} + rcos(\alpha)$$

or, replacing α with its definition,

$$l_2 = \sqrt{1 - r^2 sin^2(\pi - \theta)} + rcos(\pi - \theta), \quad r > 0, \quad \theta > \frac{\pi}{2}. \tag{10.2.8}$$

Notice that for $\theta = \pi$ we have the gamma-ray shooting left along the lower horizontal edge of the semicircle, entering the semicircle at $x = r$ and leaving the semicircle at $x = -1$, for a path length of $1 + r$, which is just what we see from Figure 10.1.1 is the correct value.

To complete this section, all we have left to do are the calculations of l_1 and l_2 for the $r < 0$ case. If you notice the symmetry of the semicircle, however, you see that in fact we don't have to do anything! All that happens when we go from $r > 0$ to $r < 0$ is flip right and left—the *geometry* of the paths is unchanged. So, we need only to randomly pick values of r from 0 to 1, and not from -1 to 1.

10.3 The Code

The code **paths.m** sends ten million random gamma-rays across a semicircle and uses (10.2.5) and (10.2.8) to compute the length of each gamma-ray path. From the vector *length* the code then creates a histogram (Figure 10.3.1) of the path lengths, from which it is seen that there is a clear difference between the number of "short" paths and the number of "long" paths.

```
%paths.m
decision=pi/2;short=0;
for path=1:10000000
    r=rand;angle=pi*rand;
    if angle<decision
        length(path)=sqrt(1-(r*sin(angle))^2)-
        r*cos(angle);
    else
        angle=pi-angle;
        length(path)=sqrt(1-(r*sin
        (angle))^2)+r*cos(angle);
    end
```

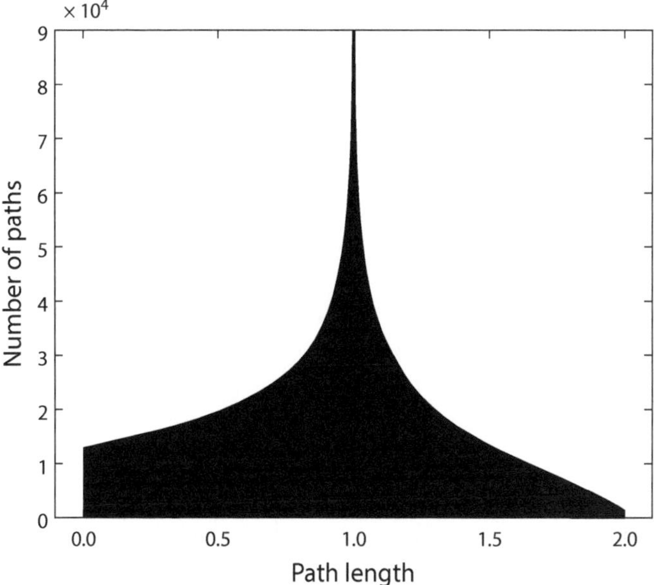

FIGURE 10.3.1. Histogram of the vector *length* in **paths.m**.

```
        if length(path)<1
            short=short+1;
        end
    end
    histogram(length,500)
    xlabel('path length')
    ylabel('number of paths')
    short/path
```

To quantify that imbalance between short and long paths, the variable *short* keeps track of the number of paths across a semicircle[4] that are shorter than 1 (a test value that is easily changed, as desired), with the result being that 58 percent of the paths are shorter than 1. If the test value is changed to 0.5, the code states that 19 percent of the paths are shorter than 0.5.

10.4 A Final Word from the Author

If only he'd paid more attention to mathematics in school.

—LAMENT OF ALBERT ROSSI, A FORMER DISHWASHER WHO HAS DEVELOPED THE ABILITY TO TIME TRAVEL BY WILL ALONE, WHILE BEING UNABLE TO UNDERSTAND THE SCIENCE BEHIND HIS REMARKABLE TALENT[5]

This book opened with the first chapter devoted to a pure number-crunching exercise involving chess matches, and so here with gamma-rays across a semicircle we have gone full circle with another number cruncher. In between, you saw some tricks (and maybe even some new math, too) on the use of a computer to tackle tough math problems that were more than number crunchers. A computer will *not,* of course, answer every math problem you may encounter for which your math ability is not quite up to what is needed—but often a computer *can* be the difference between success and failure.

Such situations will happen to everyone, sometime, no matter how many math classes they have taken. Even the great Einstein was not immune from this sad fact of life. While he understood the physical ideas behind general relativity, he lacked the mathematics to express that physics. He had to ask a mathematician friend for help, who introduced him to the new (for Einstein) mathematics of tensor calculus. Now Einstein's particular case almost certainly would not have been helped by access to an electronic computer. Yet I think his reported words, the day before he died, when he realized the goal of a grand unification of gravity, electrodynamics, and quantum mechanics would not be his, make my point here: "If only I had more mathematics."

So, there is no dodging it. There will *always* be problems that initially stop you in your tracks, and to which your first reaction might be *I am a walking dead person.* But don't despair—at least, not right away. Instead, fire up your laptop, open this book, and reread its most inspirational (for you) pages to recharge your creative energy—and write a simulation code!

APPENDIX

Got a Math Question? Try Looking Here

> Computer science . . . turns to mathematics in much the same way that engineering always has. It freely borrows from already-existing mathematics developed for altogether different purposes or, more likely, for no purpose at all. Computer scientists raid the coffers of mathematical logic, probability, statistics, the theory of algorithms, and even geometry . . . the rebirth of Euclidean geometry can be traced to the requirements of computer graphics [recall note 4 in the Prelude].
>
> —NICHOLAS METROPOLIS,[1] WHO OFTEN APPEARS IN THIS BOOK

A.1 Introduction (De Moivre Takes the Guessing Out of Probability)

You will recall that in the Prelude I wrote, "There is nothing in this book that attentive high school students who have taken an AP-calculus or AP-statistics class will find beyond them." With that claim I meant that a lot of the mathematics in this book will be stuff you have seen before in a purely *mathematical* setting. There is one particular topic, however—probability[2]—that may well represent the extension of known (to you) mathematical *ideas* into areas *not* known to you. This Appendix may help fill in that missing knowledge but, as I originally claimed, the *mathematics* itself should be nothing new.

This Appendix is not intended to be a tutorial on probability theory, but rather is best described as an admittedly often rambling collection of "thoughts on topics that occur, here and there, in the various problems discussed in this book." The whole point of this book, after all, is to show you how some mathematical questions, for which you don't have enough mathematics, may still be studied if you have a computer. Most of the theoretical analyses are here *only* to convince you that the computer solutions are correct (the last section of chapter 8 is an obvious exception). If, in those theoretical analyses, you encounter something that stops you in your tracks, then turn to this Appendix to see if the resolution to your puzzlement is here. This Appendix is not big on theorem/proof presentations, and I have taken a *lot* for granted as being intuitively clear, or at least plausible. Pure mathematicians will almost surely be appalled! Here is one example of what I am talking about.

Crucial to many of the computer codes in this book is the idea of a random number generator. In the problems themselves I have taken for granted a function available (in MATLAB it is called *rand*) that, each time you invoke it, returns a number between 0 and 1 that appears to come from a *uniform distribution* (that is, is equally likely to come from one part of the unit interval as from any other part of equal length). I will tell you a little bit of the history of such generators, and include some discussion of how such a generator "works," but it is all pretty elementary and very casual.

Remember, the *only* purpose of most of the theoretical calculations in this book is to arrive at a result that we can use to confirm the operation of a computer code. So—we begin. In the Preface (written in 1717) to his masterpiece *Doctrine of Chances*, French-born English mathematician Abraham de Moivre (1667–1754)[3] wrote of the "charm" that questions of a probabilistic nature have (as well as to make a sales pitch for his book!) as follows:

> [S]ome of the Problems about Chance having a great appearance of Simplicity, the Mind is easily drawn into a belief that their Solution may be attained by the mere Strength of natural good Sense, which generally proving otherwise, and the Mistakes occasioned thereby being not infrequent, 'tis presumed that a Book of this Kind, which teaches to distinguish Truth from what seems to nearly resemble it, will be looked upon as a help to good Reasoning.

De Moivre then went on to give two specific examples of what he had in mind:

> Among the several Mistakes that are committed about Chance, one of the most common and least suspected, is that which relates to Lotteries. Thus, supposing a Lottery wherein the proportion of the Blanks to the Prizes is as five to one, 'tis very natural to conclude, that therefore five Tickets are required for the Chance of a Prize,[4] and yet it may be proved, Demonstratively, that four Tickets are more than sufficient for that purpose, which will be confirmed by often repeated Experience. In the like manner, supposing a Lottery wherein the proportion of the Blanks to the Prizes is as Thirty-nine to One (such as was the Lottery of 1710) it may be proved, that in twenty-eight Tickets, a Prize is as likely to be taken as not; which tho' it may seem to contradict the common Notions, is nevertheless grounded upon infallible Demonstration.

To understand how de Moivre arrived at these conclusions (which are, in fact, correct), let me now introduce the concept of a *discrete random variable.*

A.2 Discrete Random Variables (Selected Topics)

The term *random variable* is the name attached to the values obtained when we measure some quantity that, from measurement to measurement, varies randomly. We will write *X* for that sequence of values. If, every time we measure X, we get a value that comes from a finite or *countably infinite* set of possible values, we say X is discrete. For example, suppose X is the result of flipping a coin that has probability p of showing heads and probability q of showing tails. Since the result of a flip is one or the other (we ignore landing on an edge!), we know that the sum of p and q must be 1 (we don't know what the result of a flip will be, but we do know it will be one or the other of the two possible results).[5] As another example, if we toss a single six-sided die, we don't know which face will show, but we *do* know one of the six possibilities *will* show. So, $\sum_{i=1}^{6} Prob(\mathbf{X} = j) = 1$. In the first example, if $p = q = 1/2$ we say the coin is *fair*, and in the second if $Prob(X = j) = 1/6$ for $j = 1, 2, 3, 4, 5, 6$ we say the die is *fair*. *Fair* means all the possibilities for the value of X are equally likely.

This all looks pretty elementary (and it is!), but with just one more idea—that of independence—we can compute the answers to some quite interesting, not-so-simple questions. We say two discrete random variables X and Y are *independent* if

$$\begin{aligned} Prob(X = j \text{ and } Y = k) &= Prob(X = j, Y = k) \\ &= Prob(X = j)\ Prob(Y = k). \end{aligned}$$

That is, if X and Y are the results of two consecutive flips of a coin (or two consecutive tosses of a die), then we say X and Y are independent if we can *multiply* the individual probabilities. So, for example, if we toss a fair die twice, the probability that the first toss shows 3 and the second toss shows 5 is $\left(\frac{1}{6}\right)\left(\frac{1}{6}\right) = \frac{1}{36}$, Similarly, the probability the sum of the two tosses is 5 is

$$\begin{aligned} Prob(X + Y = 5) &= Prob(\mathbf{X} = 2, Y = 3) + Prob(X = 3, Y = 2) \\ &\quad + Prob(\mathbf{X} = 4, Y = 1) + Prob(X = 1, Y = 4) \\ &= \left(\frac{1}{6}\right)\left(\frac{1}{6}\right) + \left(\frac{1}{6}\right)\left(\frac{1}{6}\right) + \left(\frac{1}{6}\right)\left(\frac{1}{6}\right) + \left(\frac{1}{6}\right)\left(\frac{1}{6}\right) \\ &= 4\left(\frac{1}{36}\right) = \frac{1}{9} \end{aligned}$$

and the probability the product of the two tosses is 4 is

$$Prob(XY=4)=Prob(X=1,Y=4)+Prob(X=4,Y=1)$$
$$+Prob(X=2,Y=2)$$
$$=\left(\frac{1}{6}\right)\left(\frac{1}{6}\right)+\left(\frac{1}{6}\right)\left(\frac{1}{6}\right)+\left(\frac{1}{6}\right)\left(\frac{1}{6}\right)=3\left(\frac{1}{36}\right)=\frac{1}{12}.$$

With these very simple ideas we can already do a most interesting calculation that answers the question *is there really such a thing as a perfectly fair coin*? After all, can a perfectly balanced coin, with equal probabilities of showing heads and tails, actually be made? The (perhaps) surprising answer is *yes*, a perfectly fair coin can be made and, in fact, if you have some change in your pocket, you possess such a coin right now! Here is why. Take a beat-up penny (or nickel or dime or quarter) out of your pocket: as it stands, it is certainly *not* fair and when flipped it will show heads with probability p and tails with probability q, $p \neq q$. Now, flip it twice, and so there are four possible outcomes: heads-heads (probability p^2), tails-tails (probability q^2), heads-tails (probability pq), and tails-heads (probability qp). Notice that the last two possibilities have the *same exact probability* no matter what p and q are. So, take the coin and flip it twice. If one of the first two possibilities occurs, ignore it and flip twice again. If heads-tails occurs call that Heads, and if tails-heads occurs call that Tails. Heads and Tails occur with *equal* probability and so you now have a fair coin. The price you pay for the "fairness" is having to flip multiple times to get a final result.

Here is another quite interesting calculation. Think of a positive integer from 1 to 6 and write it down on a slip of paper. Then, toss three fair dice. What is the probability at least one of the dice shows the number on the slip of paper? It's easy to get tangled up in your thinking about this, but here is a slick way to reason. There are just two possible results of the toss: at least one of the dice matches the slip, or none of the dice match the slip. That is,

$$Prob(\text{at least one die matches})+Prob(\text{none of the dice match})=1$$

and so

$$Prob(\text{at least one die matches}) = 1 - Prob(\text{none of the dice match}).$$

The probability term at the far right is easy to calculate. For each of the three dice, any face not equal to the number on the slip can show. That happens with probability $\frac{5}{6}$. So, for all three dice not to have a match, we have

$$Prob(\text{none of the dice match}) = \left(\frac{5}{6}\right)^3$$

and so

$$Prob(\text{at least one die matches}) = 1 - \left(\frac{5}{6}\right)^3 = 0.4213.$$

Do you think doubling the number of dice, from 3 to 6, or tripling the number of dice from 3 to 9, should double (or triple) the probability? There is no need to guess, but instead just calculate the two answers to be

$$Prob(\text{at least one die matches}) = 1 - \left(\frac{5}{6}\right)^6 = 0.6651$$

and

$$Prob(\text{at least one die matches}) = 1 - \left(\frac{5}{6}\right)^9 = 0.8062.$$

Are you surprised that even with nine dice (which seems like a lot) there is still a probability of almost 0.2 of there being no match?

We now have all we need to confirm de Moivre's examples. For the case of 5 blanks for each prizewinning ticket, the probability a random ticket wins a prize is $\frac{1}{6}$. Thus, the probability of that ticket *not* winning is $\frac{5}{6}$, and so n tickets fail to win a prize with probability $\left(\frac{5}{6}\right)^n$. That means the probability of winning at least one prize is $1-\left(\frac{5}{6}\right)^n$, and we wish to calculate the first value of n so that this probability is at least $\frac{1}{2}$. That is, what is the first value of n such that $1-\left(\frac{5}{6}\right)^n \geq \frac{1}{2}$? If $n = 3$, $1-\left(\frac{5}{6}\right)^3 = \frac{91}{216} < \frac{1}{2}$, while if $n = 4$,

$1-\left(\frac{5}{6}\right)^3=\frac{671}{1296}>\frac{1}{2}$, and so $n=4$ and de Moivre is correct. For de Moivre's example of the 1710 Lottery, with 39 blank tickets for every prize ticket, a ticket selected at random has probability $\frac{1}{40}$ of winning a prize. So, if you buy n tickets, you *don't* win a prize with probability $\left(\frac{39}{40}\right)^n$. In other words, you *do* win at least one prize with probability $1-\left(\frac{39}{40}\right)^n$. If $n=27$ this probability is $1-\left(\frac{39}{40}\right)^{27}=0.4951$, while if $n=28$ this probability is $1-\left(\frac{39}{40}\right)^{28}$ $=0.5078$,and we see that de Moivre's answer of $n=28$ is, again, correct.

To end this section on discrete random variables, I will show you simple derivations of two important formulas in probability theory that are used in this book. The derivations for the continuous case are a bit more subtle, but the discrete derivations are almost intuitive. So, to begin, imagine that the discrete random variable X takes on values from the set $x_1, x_2, x_3, \ldots$. Suppose we measure X a large number (N) of times, and get $X=x_1$ a total of n_1 times, $X=x_2$ a total of n_2 times, and so on. (Obviously, $n_1+n_2+\ldots=N$.) Intuitively, we write

$$Prob(X=x_1)=\frac{n_1}{N}, Prob(X=x_2)=\frac{n_2}{N}, \ldots.$$

Now, the *average* or *expected* value of X, written as $E(X)$, is

$$E(X)=\frac{n_1x_1+n_2x_2+\ldots}{N}=x_1\frac{n_1}{N}+x_2\frac{n_2}{N}+\ldots$$
$$=x_1Prob(X=x_1)+x_2Prob(X=x_2)+\ldots.$$

That is,

$$E(\boldsymbol{X})=\sum_{k=1}^{\infty}x_kProb(\boldsymbol{X}=x_k). \qquad \text{(A.2.1)}$$

That's it, *for now*. I will refer you back to (A.2.1) in the next section.

For our second formula, let me introduce you to the idea of *conditional probability*. Again, our analysis will be for the discrete case, with its primary function to eventually provide motivation for the continuous case. Imagine what mathematicians call an "experiment," with a random outcome each time it is performed, being done over and over, with a record kept of each specific outcome. For example, suppose the experiment is the simultaneous flipping of two fair coins, and so there are four possible outcomes (HH, HT, TH, TT) that form the *sample space* of the experiment, with four *sample points* of equal probability $\frac{1}{4}$. On this sample space we define two events A and B in some way. For example, A could be "two heads occurred," and B could be "at least one head occurred."

So, suppose someone flips the two coins and tells us that event B occurred. What is the probability event A also occurred? This question is written as "What is $Prob(A|B)$, the probability of A *given* B?" Before being told that B had occurred, the probability of A was $\frac{1}{4}$. Getting the additional information about B, however, changes things. Here's how to calculate *how* things change.

When we performed a large number (say, N) of double flips, we observed A alone occurred N_A times, B alone occurred N_B times, and A and B, together, occurred N_{AB} times. We would then write, intuitively,

$$Prob(A)=\frac{N_A}{N},\ Prob(B)=\frac{N_B}{N},\ Prob(AB)=\frac{N_{AB}}{N},$$

and also write

$$Prob(A|B)=\frac{N_{AB}}{N_B}$$

because this last expression is the fraction of double flips in which A occurred *on the flips that had B occurring* (remember, B is *given*). Now, notice that

$$Prob(A|B)=\frac{N_{AB}/N}{N_B/N}=\frac{Prob(AB)}{Prob(B)}. \qquad \text{(A.2.2)}$$

This expression is the result I want you to remember in the next section, but just to be complete, what *is* $Prob(A|B)$ for the coin-flipping problem? For the events A and B that we defined, notice that $Prob(AB) = Prob(A)$ and so (A.2.2) becomes

$$Prob(A\,|\,B) = \frac{Prob(A)}{Prob(B)} = \frac{1/4}{3/4} = \frac{1}{3}.$$

A.3 Continuous Random Variables (Selected Topics)

In this section we will expand our treatment of random variables to the case where the value of X, at each measurement, comes from a *continuum* (an *uncountable* infinity) of possibilities. (For example, suppose the values of X are real numbers from the interval 0 to 1.) To describe the probabilistic nature of such an X, mathematicians define a function $f_X(x)$, called the *probability density function* (pdf), such that

$$Prob(a \le \mathbf{X} \le b) = \int_a^b f_X(x)dx.$$

The pdf of X (*any* continuous X) possesses two general properties. First, even though the next value of X is unknown *until* we measure it, we *do* know that we will get *some* value and so

$$\int_{-\infty}^{\infty} f_X(x)dx = 1. \qquad \text{(A.3.1)}$$

Second, since there is no such thing as a negative probability, then it must be true that for any a and $b \ge a$ we can write

$$Prob(a \le \mathbf{X} \le b) = \int_a^b f_X(x)dx \ge 0$$

which implies, since a and b are arbitrary, that

$$f_X(x) \ge 0. \qquad \text{(A.3.2)}$$

There is an infinity of functions that satisfy these two properties, but two are especially important in computer simulation work: the *uniform* pdf and the *normal* pdf (both occur in this book). The uniform pdf is a constant over the finite interval $a \leq x \leq b$, that is,

$$f_X(x) = \begin{cases} \dfrac{1}{b-a}, & a \leq x \leq b \\ 0, & otherwise \end{cases}. \qquad \text{(A.3.3)}$$

The normal pdf has two parameters, m and σ, and is written as

$$f_X(x) = \frac{1}{\sqrt{2\pi\sigma}} e^{-\frac{1}{2}\left(\frac{x-m}{\sigma}\right)^2}, \quad -\infty < x < \infty. \qquad \text{(A.3.4)}$$

The normal pdf is a bell-shaped function that peaks at $x = m$ (the *mean*). The "width" of the bell is determined by σ (the *standard deviation*). See Figure A.4.3.

If we have more than one random variable in our problem (say, X and Y), we can talk of a *joint* pdf $f_{X,\,Y}(x, y)$ such that

$$Prob(a \leq \mathbf{X} \leq b,\, c \leq \mathbf{Y} \leq d) = \int_c^d \int_a^b f_{X,Y}(x,y)dxdy, \qquad \text{(A.3.5)}$$

that is, the probability of interest is the integral of the joint pdf over an area patch defined by the limits of integration. If X and Y are *independent*, then

$$\begin{aligned} &Prob(a \leq X \leq b,\ c \leq Y \leq d) \\ &\quad = Prob(a \leq X \leq b)Prob(c \leq Y \leq d) \\ &\quad = \int_a^b f_X(x)dx \int_c^d f_Y(y)dy \\ &\quad = \int_c^d \int_a^b f_X(x) f_Y(y)dxdy\,. \end{aligned} \qquad \text{(A.3.6)}$$

So, if X and Y are independent we see, by comparing the double integrals at the far right of (A.3.5) and (A.3.6) and recalling that a, b, c, and d are arbitrary, that the joint pdf is just the product of the individual pdfs. That is, if X and Y are independent

$$f_{X,\,Y}(x, y) = f_X(x) f_Y(y)$$

Notice that if X and Y are independent and each is uniform from 0 to 1, then

$$f_X(x)=\begin{cases} 1, & 0\leq x\leq 1 \\ 0, & otherwise \end{cases}$$

and

$$f_Y(y)=\begin{cases} 1, & 0\leq y\leq 1 \\ 0, & otherwise \end{cases}$$

and so

$$f_{X,Y}(x,y)=\begin{cases} 1, & 0\leq x,y\leq 1 \\ 0, & otherwise \end{cases}.$$

Thus, if X and Y are independent and each is uniform over the unit square, we have

$$Prob(a\leq \boldsymbol{X}\leq b,\ c\leq \boldsymbol{Y}\leq d)=\int_c^d\int_a^b dxdy=(b-a)(d-c)$$

which is the *area* of a patch in the unit square defined by the limits of integration.

Mathematicians use the pdf to define a new function called the *distribution*, $F_X(x)$, written as

$$F_X(x)=Prob(\boldsymbol{X}\leq x)=\int_{-\infty}^x f_X(u)du. \qquad \text{(A.3.7)}$$

Since $f_X(x)\geq 0$ (recall (A.3.2)) then as x increases so does $F_X(x)$ because the integration operation in (A.3.7) picks up ever more of the area under the pdf curve. That is, a distribution function *monotonically increases with increasing argument*, starting from $F_X(-\infty)=0$ and steadily increases to $F_X(\infty)=1$. Another way to say this is that if $a<b$ it then follows that $F_X(a)<F_X(b)$ or, going in the other direction, if $F_X(a)<F_X(b)$ it then follows that $a<b$. I will remind you of this in the next section when we get to random number generators. Finally, since differentiation is the "inverse" operation to integration, we can write

$$f_X(x)=\frac{d}{dx}F_X(x). \qquad \text{(A.3.8)}$$

The Los Alamos bomb analysts used the distribution function concept in support of their computer studies of neutron chain reactions in uranium-235 (at the end of the next section I will say more on that). But first, here is a simple example of how the distribution function works *mathematically*. Suppose X and Y are independent random variables, each uniformly distributed on the interval 0 to 1. That is, $f_X(x)=1, f_Y(y)=1$, and so the joint pdf is $f_{X,Y}(x, y)=f_X(x) f_Y(y)=1$ over the unit square $0 \leq x, y \leq 1$. That square, which contains all the possible values of X and Y, is called the *sample space*. Now, suppose we define a third random variable $Z=X+Y$ and ask: what is the pdf of **Z**? We know Z will clearly take on values from the interval 0 to 2, but with what probability density (as you will see, the pdf of Z is *not* uniform)? To answer this question, we start by calculating the distribution of Z. That is, we will first calculate

$$F_Z(z)=\int_0^z f_Z(u)du$$

and then differentiate to write

$$f_Z(z)=\frac{d}{dz}\int_0^z f_Z(u)du.$$

We have, where z is some value from 0 to 2,

$$F_Z(z)=Prob(Z \leq z)=Prob(X+Y \leq z)=Prob(Y \leq z-X).$$

This last probability is most easily visualized with the aid of Figure A.3.1, which shows the line $Y=z-X$ plotted on the unit square, where z is taken to be some value between 0 and 1 in the top half of the figure, and some value between 1 and 2 in the bottom half. The probability we are after is, as mentioned earlier, the shaded area in the two halves, the area *below* the $Y=z-X$ line (*below* because of the *less*-than-or-equal inequality in the distribution function argument).

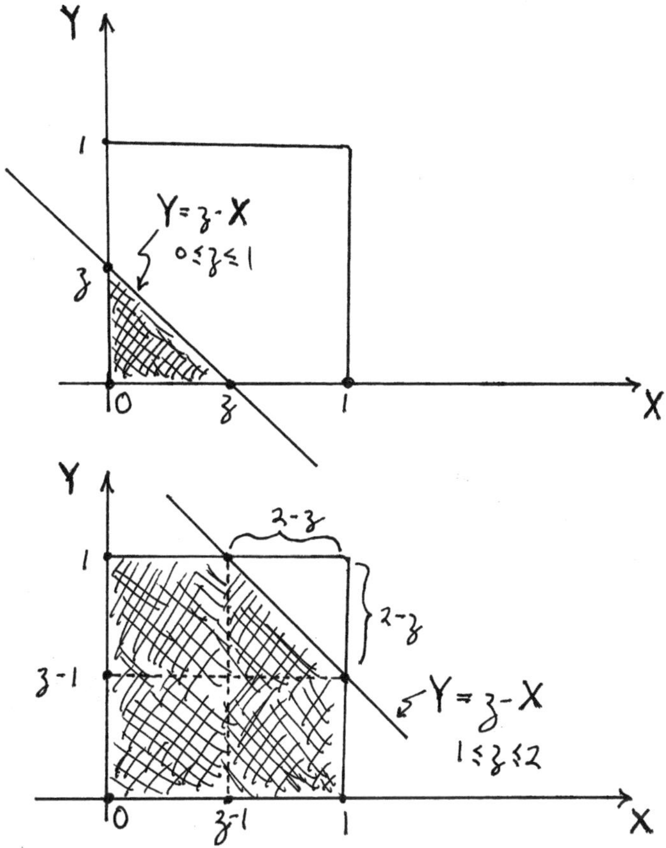

FIGURE A.3.1. Calculating the distribution function of $\boldsymbol{Z}=\boldsymbol{X}+\boldsymbol{Y}$ when both $\boldsymbol{X}$ and $\boldsymbol{Y}$ are uniform from 0 to 1.

Thus,

$$F_Z(z)=\begin{cases} \frac{1}{2}z^2, & 0\leq z\leq 1 \\ 1-\frac{1}{2}(2-z)^2. & 1\leq z\leq 2 \end{cases},$$

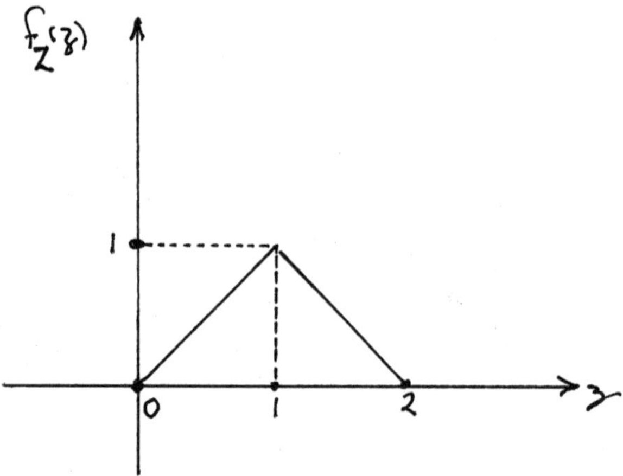

FIGURE A.3.2. The pdf of **Z** when both **X** and **Y** are uniform from 0 to 1.

and so

$$f_Z(z) = \begin{cases} z, & 0 \le z \le 1 \\ 2 - z, & 1 \le z \le 2 \end{cases}.$$

Thus, as shown in Figure A.3.2, the pdf of $Z = X + Y$ is *triangular*.

Notice that $f_Z(z)$ satisfies both of the general properties of (A.3.1) and (A.3.2) that hold for *any* pdf. We can experimentally verify this theoretical calculation by actually generating a lot of pairs of values for X and Y, summing the pairs, and plotting a histogram of the sums. Figure A.3.3 shows the result of generating ten million such pairs (see the following box) using MATLAB's uniform random number generator (more about that in the next section), and the result does indeed look just like Figure A.3.2.

```
%sumof2rv.m
for loop=1:10000000
    x=rand;y=rand;
    z(loop)=x+y;
end
histogram(z,500)
```

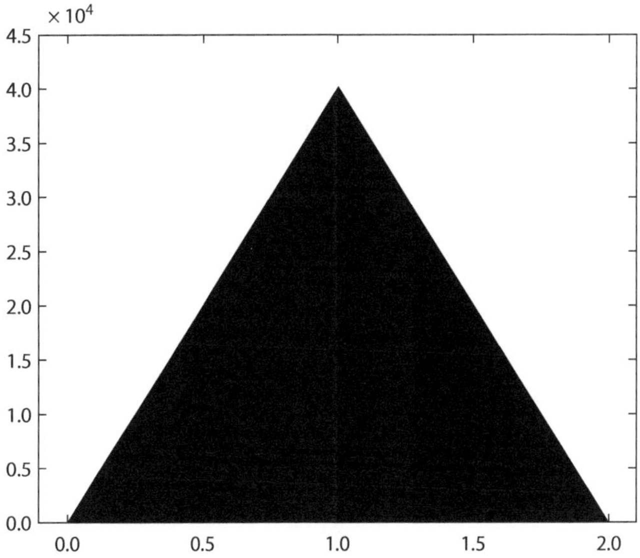

FIGURE A.3.3. Experimental verification of the calculated $f_Z(z)$.

To finish this section, first recall our result (A.2.1) for the expected value of a discrete random variable: $E(\mathbf{X})=\sum_{k=1}^{\infty} x_k \, Prob(\mathbf{X}=x_k)$. For the continuous case, the summation becomes an integral and $Prob(X=x_k)$ becomes $f_X(x)dx$. Thus,

$$E(\boldsymbol{X})=\int_{-\infty}^{\infty} x f_X(x)dx.$$

Finally, recall our result (A.2.2) when discussing conditional probability in the case of discrete random variables, with A and B as *events* defined on the same sample space:

$$Prob(A\,|\,B)=\frac{Prob(AB)}{Prob(B)}.$$

For continuous random variables, the analogous formula for pdfs relates the *conditional* pdf $f_{X|Y}(x, y)$ to the joint pdf of X and Y—$f_{X,Y}(x, y)$—and the pdf of Y, $f_Y(y)$. That is,

$$f_{X|Y}(x, y) = \frac{f_{X,Y}(x, y)}{f_Y(y)}.$$

In introductory probability theory books this is usually simply given as a definition because its formal derivation is not elementary. We can easily verify that $f_{X|Y}(x, y)$ is indeed a pdf, however, and so the definition is at least plausible. First, since both $f_{X,Y}(x, y)$ and $f_Y(y)$ are nonnegative, then their ratio is, too. Also, if we integrate x out of both sides,

$$\begin{aligned}\int_{-\infty}^{\infty} f_{X|Y}(x, y)dx &= \int_{-\infty}^{\infty} \frac{f_{X,Y}(x, y)}{f_Y(y)} dx \\ &= \frac{1}{f_Y(y)} \int_{-\infty}^{\infty} f_{X,Y}(x, y)dx \\ &= \frac{1}{f_Y(y)} f_Y(y) = 1.\end{aligned}$$

A.4 Random Number Generators (a *Very* Casual Overview)

In this section I will take it for granted you are already well aware of the use of random numbers in the sort of computer problems treated in this book. Long before the coming of high-speed electronic computers the importance of random numbers was appreciated, but the invention of electronic computation suddenly added a whole new dimension to the subject—the question of how to get a *lot* of random numbers in a very short time interval. Say, for example, a million numbers every second, for perhaps minutes, hours, or even days at a stretch. The earliest known technique, tossing dice (more generally, "casting lots") for generating random numbers was, alas, a *lot* slower than that. But it is an *old* technique, having been traced back to 2,500 years *before* Christ.

In fact, the state of the art remained pretty much unchanged even as recently as the end of the nineteenth century, 2,000 years *after* Christ. That's when we find British statistician Francis Galton (1822–1911) waxing poetically about dice, as follows:

> As an instrument for selecting at random, I have found nothing superior to dice. It is most tedious to shuffle cards . . . and the method of mixing and stirring-up marked balls in a bag is more tedious still. . . . [S]ome form of roulette is preferable to these, but dice are better than all. When they are shaken and tossed in a basket, they hurtle so variously against one another and against the ribs of the basket-work that they tumble wildly about, and their positions at the outset afford no perceptible clue to what they will be after even a single good shake and toss.[6]

Ten years later, however, we find Lord Kelvin, despite Galton's rejection of card shuffling, having to use that very method (along with coin flipping) to generate random numbers for a Monte Carlo simulation of a gas (look back at the second paragraph of chapter 2). To generate the paths of gas molecules as they moved through *thousands* of interactions with other molecules and confining surfaces, the (random) outcome of *each* such interaction was determined by a shuffle of 100 cards and the flipping of a coin[7] to get a total of 200 possible random numbers. Doing that was such an awful task that Kelvin felt obligated to immediately open his paper by offering thanks to his assistant (who did the actual shuffling and flipping) for performing the tedious work with "unfailing faithful perseverance."

To find such a dedicated assistant is not easy, so an alternative approach was clearly needed for future simulations. One interesting idea is to somehow use a physical process that is inherently random: one example of that is radioactive decay or, as von Neumann put it, "nuclear accidents."[8] He immediately appreciated, however, a fatal flaw in such an approach: "The real objection to this procedure is the practical need for checking computations. If we suspect that a calculation is wrong, almost any reasonable check involves repeating something done before. At that point the introduction of new random numbers would be intolerable. *I think that the direct use of a physical supply of random digits is absolutely inacceptable for this reason and this reason alone* [my emphasis]." A necessity was a way to generate the *same* random numbers, over and over, a requirement that appears to be a self-contradiction. What we are after is a way to generate, *algorithmically,* what *seem* to be random numbers (that is, *pseudo*-random numbers). Thinking about this prompted von Neumann to issue this famous warning: "Any one who considers arithmetical methods of producing random digits is, of course, in a state of sin. . . . There is no such thing as a random number—there are only methods to produce random numbers, and a strict arithmetic procedure of course is not such a method [note 8]."

With no little irony, the appearance of the high-speed electronic computer, with its voracious, seemingly insatiable appetite for random numbers when doing Monte Carlo simulations, also provided the solution to von Neumann's quandary. That's because it was soon realized that the computer *itself* could be used to generate the pseudo-random numbers it so eagerly consumed. It is probably no great surprise that von Neumann put forth the first idea for how that could be done;[9] it proved to not be a very good idea, but it *was* a start, and it *was* used for years on the Los Alamos MANIAC.

Von Neumann's idea, called the *middle-square method,* does, at first glance, appear to combine promise with simplicity. To illustrate it, start with a four-digit number, say $x_0 = 2061$. (This starting number is called the *seed.*) Square it to get 0424771 and take the middle four digits to write $x_1 = 2477$. Square it to get 0613559 and take the middle four digits to write $x_2 = 1355$. Keep doing this to get $x_3 = 8360$, $x_4 = 8890$, and so on. Finally, divide each of these numbers by 10,000 to get numbers in the interval 0 to 1. The ultimate goal of all the random number generators that I am aware of in modern-day computer software is that of behaving as a *uniform* generator from 0 to 1. That's because, as I will show you by the end of this section, from a uniform 0-to-1 distribution it is theoretically possible to generate random numbers that have *any other* distribution that you might desire.

The middle-square process is easy to program (it was first used on the ENIAC), and the hope was that the resulting numbers would, in fact, be a uniform distribution from 0 to 1. One immediate problem with the middle-square method is that the numbers generated are going to eventually repeat (there are, after all, only 10,000 four-digit numbers and, once a number repeats, the sequence up to that number repeats[10]). Despite that, and other more subtle problems, the middle-square method was indeed used on the Los Alamos MANIAC for years (with numbers larger than four digits).[11]

In addition to having a long period, a good random number generator must "pass" various statistical tests to be declared *good.* The most obvious is simply a frequency test that checks to see if the digits 0 to 9 each appear one-tenth of the time, if the pairs 00 to 99 each appear one-hundredth of the time, if the triplets 000 to 999 each appear one-thousandth of the time, and so on. Numerous other tests look to see if more subtle statistical anomalies are present. Von Neumann called (note 8) arithmetic methods for generating random numbers "mere 'cooking recipes' for making digits . . . some statistical study of the digits generated by a given recipe should be made, but exhaustive tests are impractical."

A completely different approach to generating random numbers came in 1949, with the introduction of the *congruential generator* by American mathematician D. H. Lehmer (1905–1991). Like the middle-square method, this is a deterministic procedure for computing the next random number in a sequence from the current random number. If x_k is the kth number in a sequence, then the next number is calculated as

$$x_{k+1} = (kx_k + c) \; mod \; N$$

where k, c (often taken to be zero), and N are "appropriately chosen" constants.[12] The values of $k = 23$ and $N = 10^8 + 1$ were used, for example, on the ENIAC, to give a random number sequence with a period greater than five million.

Over the years many people spent a lot of time attempting to determine the “best” values for k, c, and N. In the late 1990s that all became moot when Japanese computer scientists discovered a nonarithmetic way to generate random number sequences uniform from 0 to 1, sequences with enormous periods[13] combined with excellent statistical behavior. Called the *Mersenne Twister* (because it is based on Mersenne primes—primes of the form $2^p - 1$ where p is, itself, a prime—and shift registers with feedback connections). Today the Twister is the uniform generator implemented in MATLAB as well as in many other scientific programming languages.

There is a *lot* of firepower behind MATLAB’s simple-looking *rand* function!

In a letter dated May 21, 1947, von Neumann wrote to mathematician Stanisław Ulam about two very different ways to generate random numbers on a computer.[14] The two methods are concerned with generating numbers that appear to come from probability distributions other than a uniform one, *from* numbers that *do* come from a uniform 0-to-1 generator. The first method is outlined as a summary of an idea that von Neumann credits Ulam for originating. I will conclude this section by telling you about both methods.

Here’s the problem that Ulam solved: Suppose X is uniform from 0 to 1, and Y (the desired random quantity) is to have some specified distribution $F_Y(y) = Prob(Y \leq y)$. How do we generate the values of Y from values of X? The answer is beautiful, both in derivation and in form. We start by observing that since X is uniform, its distribution is

$$F_X(x) = Prob(X \leq x) = x, \quad 0 \leq x \leq 1. \qquad \text{(A.4.1)}$$

Since *by definition* $F_Y(y)$ is a probability, it has a value in the interval 0 to 1 and so with (A.4.1) as a template, we have

$$Prob(X \leq F_Y(y)) = F_Y(y). \tag{A.4.2}$$

On the left-hand side of (A.4.2), if we take the inverse function of both sides of the inequality in the argument, then by the monotonicity of distributions (look back at (A.3.7)), we don't change the sense of the inequality and so arrive at

$$Prob(F_Y^{-1}(X) \leq F_Y^{-1}(F_Y(y))) = F_Y(y). \tag{A.4.3}$$

Since $F_Y^{-1}(F_Y(y)) = y$ then (A.4.3) becomes

$$Prob(F_Y^{-1}(X) \leq y) = F_Y(y) = Prob(Y \leq y). \tag{A.4.4}$$

Comparing the term to the left of the first equals sign in (A.4.4) to the term to the right of the last equals sign in (A.4.4), and remembering (A.4.4) holds for any value of y, we have Ulam's answer:

$$Y = F_Y^{-1}(X). \tag{A.4.5}$$

Here is an example of how this works. Suppose we want Y to have values that come from an exponential distribution,[15] that is, the pdf of Y is

$$f_Y(y) = \begin{cases} e^{-y}, & y \geq 0 \\ 0, & y < 0 \end{cases}.$$

So, we have

$$F_Y(y) = \int_0^y e^{-u}du = -e^{-u}\Big|_0^y = 1 - e^{-y}, \quad y \geq 0. \tag{A.4.6}$$

If we write $F_Y^{-1}(y)$ in place of y in (A.4.6)

$$F_Y(F_Y^{-1}(y)) = 1 - e^{-F_Y^{-1}(Y)}$$

or, recalling as before that $F_Y^{-1}(F_Y(y)) = y$, then

$$y = 1 - e^{-F_Y^{-1}(Y)}$$

which reduces to

$$F_Y^{-1}(y) = \ln(1-y). \tag{A.4.7}$$

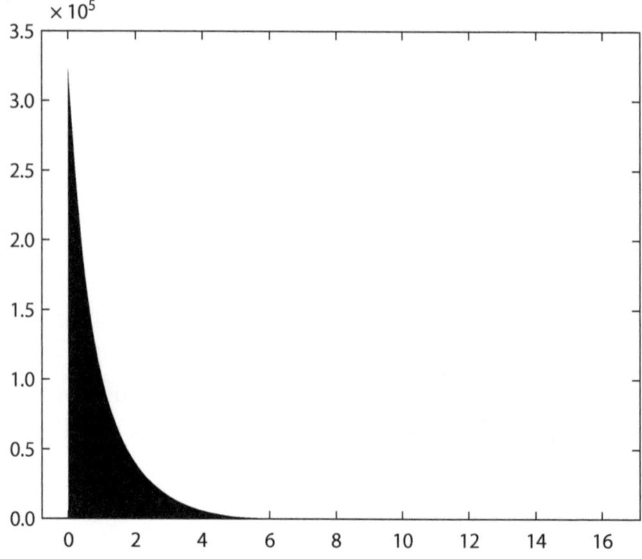

FIGURE A.4.1. Ulam's solution for an exponential pdf.

Writing X for y in (A.4.7), and recalling (A.4.5), we have Ulam's answer:

$$Y = -ln(1 - X). \tag{A.4.8}$$

In the same way that we generated Figure A.3.3 to get a histogram approximation of the pdf of the sum of two uniform independent random variables, Figure A.4.1 shows a histogram of Ulam's solution (for ten million values of Y). We see that the histogram does, indeed, look like a decreasing exponential.

Von Neumann had what I think a curious reaction to (A.4.8), saying (note 8) that while it all makes mathematical sense, nonetheless "it seems objectionable to compute a transcendental function of a random number." He didn't elaborate on *why* he felt this way (I will stick my neck out here to disagree with a genius and say I think von Neumann's concern unwarranted), but it did motivate him to search for another solution. And he found one! That was actually good, too, because it isn't always possible to use

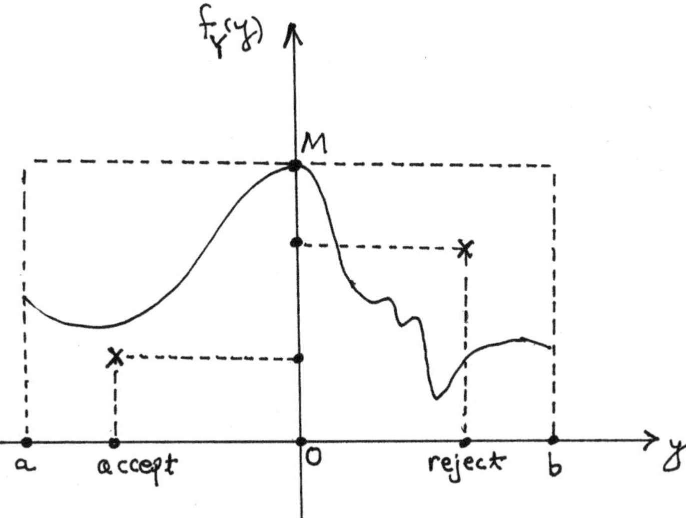

FIGURE A.4.2. Von Neumann's accept/reject method.

Ulam's theoretically perfect solution since it may just not be possible to analytically find $F_Y^{-1}(X)$. The classic example of that possibility is if Y is normal, because there is no indefinite integral of e^{-y^2}.[16] Von Neumann's alternative approach, however, does the job with ease. Here's how it works.

Imagine that the pdf of Y (the random quantity for which we wish to generate values) is as shown in Figure A.4.2. That pdf is sufficiently "awful" that it is highly unlikely we could ever find its distribution analytically (much less invert it!). *With von Neumann's approach we don't have to.* As shown in the figure, we imagine that the values of Y are in the interval $a \leq y \leq b < \infty$ and the pdf has a maximum value of M. Then, generate two independent numbers from a uniform distribution, one from the interval a to b and the other from the interval 0 to M. These two values are the coordinates of a point in the rectangular box indicated in Figure A.4.2. If that point is such that $f < f_Y(y)$ (that is, the point lies *beneath* the pdf), then *accept* y as a value for Y, but if $f > f_Y(y)$ (that is, the point lies *above* the pdf), then *reject* y as a value for Y. If we do this a large

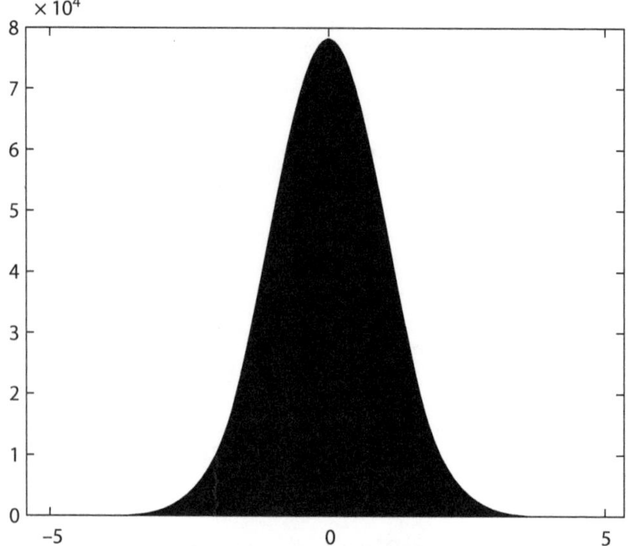

FIGURE A.4.3. Von Neumann's accept/reject method builds a normal random number generator (zero mean/unit variance) from a 0 to 1 uniform generator.

number of times, the fraction of accepted points will "mirror" the ups and downs of the pdf of Y.

Now, suppose $f_Y(y) = e^{-y^2/2}$ with $a = -5$ and $b = 5$. Obviously, $M = 1$. That is, Y is normal with zero mean and unit standard deviation, and so von Neumann's accept/reject algorithm should give us a bell-shaped histogram from -5 to 5. Does it? Here's the code **vN.m** (in honor of von Neumann) that implements the method, along with Figure A.4.3 that displays the histogram of ten million accepted values for Y.

```
%vN.m
accept=0;
while accept<10000000
    y=-5+10*rand;f=rand;
    if f<exp(-(y^2)/2)
        accept=accept+1;
```

```
            value(accept)=y;
        end
    end
    histogram(value,500)
```

I will end this section on two amusing notes, just to show that a discussion of random number generators can have its lighter moments. I start with the admission (note 8) by von Neumann (when comparing his method with Ulam's for the case of generating values for an exponential pdf): "It is a sad fact of life [that for ENIAC] it was slightly *quicker* [my emphasis] to use a truncated power series for $log(1-x)$ [that is, to implement Ulam's solution] than to carry out [the test for acceptance or rejection]." I suspect this admission was at least just a little bit painful for von Neumann to make.

Second, for your entertainment, what follows is a brief fictional (*I think*) tale of random numbers in a theological setting. It was published by *Analog Science Fiction Magazine* (the renamed *Astounding Science Fiction Magazine*) in 1984, under the title of "Some Things Just Have to Be Done by Hand!" but, if submitting it today, I would instead use what I think the better title of

"Random Numbers in Heaven"

The Most Important Entity rubbed His temples in fatigue. There was just so damned much crap to put up with nowadays. The personnel paperwork was nearly overwhelming, even for a being with omnipotent powers. And a work force faced with zero turnover had a first-class morale problem. The younger ones knew there was no hope for advancement by the once-usual routes of death, retirement, or resignation. None of those events ever happened—here. The telephone rang, and He answered in weary relief at the distraction.

"Yes?"

"Sorry to bother you, Sir, but the main computers have a backlog in the RANDOM QUEUE for ten to the 183rd power decisions. Can you please service those requests right now?"

"Random Numbers in Heaven" (Continued)

"Damn, are those bloody scientists on Earth doing their quantum experiments again!? You'd think they'd understand the Uncertainty Principle after all these years. Well, what is it now, an electron beam through a diffraction grating, or is somebody trying to locate an atom with zero error?"

"Both, and more, Sir. Those guys are really getting busy down there. Why, just as we've been talking here, the RQ has picked up ten to the 179th power more requests!"

The main computers couldn't be allowed to overflow. Once, two or three thousand years ago (in Earth time), they had been unattended for several days (in His time), and the RQ had clogged up tight with ignored decision requests for determining the outcomes of random events. The resulting massive computer system crash had caused entire centuries (in Earth time) of strange, abnormal violations in His Laws of Natural Phenomena. It had been the time of magic on Earth, and the new wizards, sorcerers, and magicians had used it to their advantage in proclaiming themselves to be all-powerful. It couldn't be allowed to happen again!

"All right, all right, hold your feathers smooth. Hang on for a moment."

He put His caller on hold, and pulled open the desk drawer next to His perfect left foot. Inside was a pure diamond crystal box, containing two ruby cubes of ultimate clarity. The dots on the cube faces were precise circles of gold. The cubes were perfectly balanced, of course, as it was impossible for anything unfair to exist—here. Taking the cubes in His mighty hand, He established a mind-link with the input-output data lines to the main computers. Faster than imaginable (or even possible by ordinary laws, but for Him very little was impossible) the cubes tumbled in His quivering hand. The whole thing was over in just a few wing beats. He dropped the cubes, now so hot they glowed in the gamma-ray region of the spectrum, back into their crystal box, and shoved the drawer shut with a kick from His perfect left foot.

"Okay, the main computers cleaned up?"

"Yes Sir, the RANDOM QUEUE is empty!"

"Excellent—now please don't call again for at least another day. Meanwhile, you and your colleagues might busy yourselves with finding a way to speed up the automatic software random number generator. I find this business of hand generation to be increasingly inconvenient. Good-bye."

As He hung-up, He thought of what Albert Einstein, one of the better Earth scientists, had once said: "God doesn't play dice with the Cosmos."

"Hummph," He grunted in disgust to Himself, "just what the Hell did he know about it?"

To those who might feel it just a bit outrageous to imagine a random number story set in Heaven, the very next year the radio astronomer Carl Sagan (1934–1996) published his best-selling novel *Contact* (made into a 1997 movie) that ends in an equally startling, quasi-theological fashion. After an intense examination of a computer calculation of the decimal digits of pi,[17] it is discovered that the apparent randomness[18] of those digits contains a long sub-sequence of all ones and zeros that displays as a two-dimensional image of . . . *a circle*! Holy cow! Sagan calls this "The Artist's Signature." When Sagan wrote, pi was known to more than 16 million decimal digits, and today (2024) pi is known to many *trillions* of digits, and no such picture has been discovered (yet). . . . But who knows what tomorrow may bring?[19]

As a final note the randomness of pi is made all the more mysterious by the fact it has been known, *for centuries*, that there is a definite *non*-random structure to pi. As discovered by English mathematician William Brouncker (1620–1684) in 1656 (by unknown means), we can write the beautiful continued fraction

$$\frac{4}{\pi}=1+\cfrac{1^2}{2+\cfrac{3^2}{2+\cfrac{5^2}{2+\cfrac{7^2}{2+\ldots}}}}$$

where the division bars go on forever. In 1775 Swiss-born mathematical physicist Leonhard Euler (1707–1783) discovered a derivation that is accessible to any high school student who has taken AP-calculus.[20] I think the joint truth of pi being highly structured *as well as random* is a very deep mathematical puzzle that is still not understood.

A.5 The Monte Carlo Method (Ulam's Great Idea)

In this final section of the Appendix, I will present two simple but illustrative examples of what almost this entire book is about: a method that "solves" problems even if you can't do them theoretically. This so-called *Monte Carlo method* is possible only because mathematicians discovered ways to generate essentially an endless stream of numbers at electronic speed that, statistically, appear to come from a random process. The Monte Carlo method (the name was prompted by the famous gambling casino in Monaco) is generally credited to Polish-born American mathematician Stanisław Ulam (who appears in the opening sections of this book, as well as in the previous section of this Appendix)—but the *idea* of the method had been around long before Ulam (Figure A.5.1). What Ulam brought to the table that was new was a significant problem (the fission chain reaction central to the physics of the bomb) for which Monte Carlo was the only way that would work, along with access to the high-speed electronic ENIAC (and to the Los Alamos MANIAC computer, after that).

Ulam tells readers how he came to think of the Monte Carlo idea—while playing solitaire during his recovery from a serious illness—in his autobiography[21], which is, half a century after it appeared and more than forty years after his death, still a great read. However, rather than repeat that tale here (read his book!), let me simply show you a problem that we will first solve analytically, and then again with the Monte Carlo approach. You will see just how close to theory the computer code proves to be. The problem itself is a bit on the whimsical side, but it is a popular

FIGURE A.5.1. Stanisław Ulam's WWII Los Alamos security badge photo. (Upon arrival at Los Alamos each person was assigned a unique letter/number ID that was displayed on their badge: Ulam's was H0, Metropolis' was G15, and von Neumann's was M50.) Image courtesy of The Los Alamos National Laboratory.

math puzzle that is easy to visualize and (perhaps) not so easy to solve (until you see the trick).

Imagine a wire cube, as shown in Figure A.5.2, with a spider at one vertex (S) and a fly at the body-diagonal vertex (F). (I will explain soon what the lowercase letters at each vertex denote, as well as what the binary numbers indicate.) The fly is unable to move as it is frozen in place by fear, but the spider can move along the edges of the cube. The spider is blind, however, and makes its decision on how to move from the vertex it presently occupies to

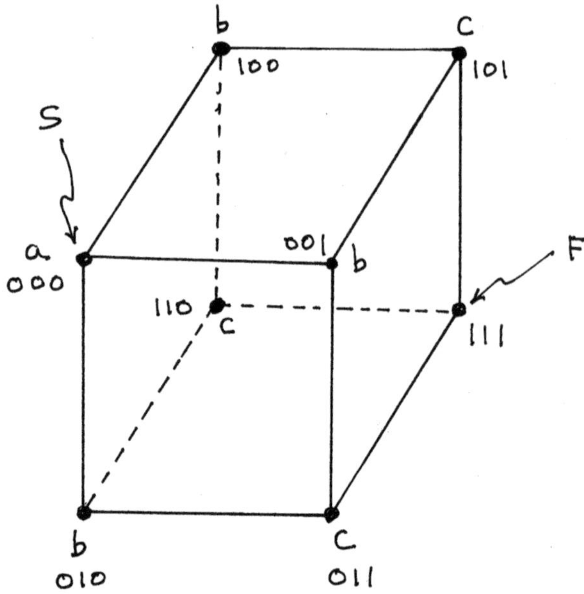

FIGURE A.5.2. The spider-and-fly problem.

an adjacent vertex at random (there are always *three* equally likely possibilities for the spider, including a return to the vertex it just left). The minimum number of steps to reach the fly is obviously 3, but a trip to the fly could take considerably more steps, depending on the spider's decisions. Our question is this—if we repeat this sad (for the fly) little story a large number of times, how many steps, *on average*, does the spider take to reach the fly?

Let's first solve the problem theoretically, with the symbol a denoting the average number of steps to reach the fly from the starting vertex. Notice that this vertex is three edges distant from the target vertex (the fly). It seems reasonable to assume that the closer a vertex is to the target vertex the smaller will be the average number of steps to reach the target vertex. So, if we write b as the average number of steps to reach the target vertex if the spider is at any vertex that is two edges distant from the target vertex (see Figure A.5.2 again), we would expect $b < a$. And finally, let's write c as the average number of steps to reach the target vertex given that the spider is at any vertex that is one edge distant from the target vertex, and we would expect $c < b$.

Now, here's the crucial observation that reduces the problem to a simple high school algebra problem. If the spider is at vertex 000 (see Figure A.5.2 again) it can move with equal probability to vertex 001, vertex 010, or vertex 100. For each of those three vertices, the average number of steps *to go from there* is b. So, since it takes 1 step to go from 000 to one of those three adjacent vertices, we can write

$$a = \frac{1}{3}(1+b) + \frac{1}{3}(1+b) + \frac{1}{3}(1+b)$$

where I have weighted each possibility by the probability of that possibility. Similarly, if the spider is at any vertex with b as the average number of steps to reach the target vertex, we can write

$$b = \frac{1}{3}(1+a) + \frac{1}{3}(1+c) + \frac{1}{3}(1+c).$$

And finally, if the spider is at any vertex with c as the average number of steps to reach the target vertex, we can write

$$c = \frac{1}{3}(1+b) + \frac{1}{3}(1+b) + \frac{1}{3}(1),$$

where the last term in this expression for c represents the fact that the spider's step from its current vertex has at last delivered it to the fly.

These three equations in three unknowns are easily solved (I will let you fill in the details), or you can simply take my word for it and just verify that the solution is $a = 10$, $b = 9$, and $c = 7$. Notice that $c < b < a$, which agrees with our physical commonsense assumptions from a few steps back. The value of $a = 10$ might be a surprise, however, perhaps looking rather large for the *average* number of steps, since that means there must be random walks on the cube that are even longer. That observation prompts this concern: is our theoretical analysis correct? To answer that, let's simulate the problem and see what the computer says.

To simulate the motion of the spider, notice that as it moves from one vertex to the next, the binary code of the new vertex is simply the binary code of the last vertex with one digit changed. So, if the spider is initially at 000, all we need do is randomly flip the digits one at a time (each flip representing a step) until we get to 111 (the target vertex). The code will do that for ten million random walks, keeping a running total of the total number of flips (steps). Dividing that final total by ten million will give us the average number of steps per walk. That's what the code **spiderfly.m** does. A final twist to doing all this is to not use 0 and 1 as the binary digits to code the vertices, but rather −1 for 0 and +1 for 1. That's convenient to do because it is easy to flip −1 and +1 (just change the *sign*). The only special MATLAB command in the code is *sum*, which adds all the elements of the three-element vector V that holds the binary code for the spider's current vertex. So, we start the spider at $V = [-1\ -1\ -1]$ and put the fly at $[1\ 1\ 1]$. When the sum of the elements of V equals 3, the code "knows" the spider has arrived at the fly.

```
%spiderfly.m
r1 = 1/3;r2 = 2/3;totalsteps=0;
for loop=1:10000000
     V(1)=-1;V(2)=-1;V(3)=-1;
     while sum(V)<3
          r=rand;
          if r<r1
               V(1)=-V(1);
          elseif r<r2
               V(2)=-V(2);
          else
               V(3)=-V(3);
          end
```

```
        totalsteps=totalsteps+1;
        end
end
totalsteps/loop
```

When run, the code declared the average number of steps to be 10.0045. And just to check a bit further, when started with $V=[1\ -1\ -1]$, for which the analysis predicted an average of 9 steps, the code states the average is 9.0028. And when started with $V=[1\ 1\ -1]$ the code states the average is 7.0014 steps. Pretty good agreement with theory!

Next, to show you the power of Monte Carlo, consider the following variant of the spider-and-fly problem. Initially, the spider and the fly can be located on the cube at any two vertices (not necessarily ones on a body diagonal). The spider is still blind and so makes its next-step decision at random, just as before, and the fly is still frozen in place with fear. What *has* changed is our question: What is the probability the spider reaches the fly before it randomly stumbles back onto its starting vertex (where it abandons the hunt in frustration)?

To theoretically analyze this problem takes a bit more math than did the previous problem, but you can find a treatment in the little monograph by Doyle and Snell (note 10 in chapter 5). On their page 58 it is shown that if initially the spider and fly are on any two vertices that are one edge apart (for example, the spider starts at 000 in Figure A.5.2 and the fly is at 010), the probability in question is $\frac{4}{7}=0.5714\ldots$. We can check that claim with the Monte Carlo code **sfv.m** (for spider-fly-variant). Here is how it works.

```
%sfv.m
r1 = 1/3;r2 = 2/3;bingo=0;
for loop=1:10000000
        stop=0;
        V(1)=-1;V(2)=-1;V(3)=-1;
        R=V;T=[-1 1 -1];
        while stop==0
```

```
            r=rand;
            if r<r1
                 V(1)=-V(1);
            elseif r<r2
                 V(2)=-V(2);
            else
                 V(3)=-V(3);
            end
            if V==R|V==T
                 if V==T
                      bingo=bingo+1;
                 end
                 stop=1;
            end
       end
  end
  bingo/loop
```

Much of the code has been lifted from **spiderfly.m**, but there have been some significant changes, too. The **sfv.m** code retains the use of ±1 binary coding for vertices, and so the spider starts at $V = [-1\ -1\ -1]$ and the fly is fixed at $T = [-1\ 1\ -1]$. The vector R is set equal to the initial value of V so the code will "know" when (if) the spider has *returned* to its starting vertex. After each step is taken, and V now records the binary code for the spider's new location, the code looks to see if V is equal to either R (and so the spider has returned) or to T (and so the spider has reached the fly). The code line *if V==R|V==T* carries out that check, where the vertical line is MATLAB's logical inclusive-OR operation. Both conditions mean the present walk is over, but only the $V = T$ condition causes *bingo* (the number of walks where the spider reaches the fly) to be incremented. Both conditions set *stop* to 1, which terminates the *while* loop and triggers the code to begin a new walk. When run, **sfv.m** reported a probability of 0.571 . . . , which is satisfyingly close to $\frac{4}{7}$.

ACKNOWLEDGMENTS

I WORKED HARD TO put this book together, but I did not work alone. I have benefited from the talents of not just a few people and institutions and I am grateful to all. At Princeton University Press, my editor, Diana Gillooly, was supportive of this project from the moment she first read my proposal, and the book has benefited hugely from her expert guidance. Diana's efficient assistant, Whitney Rauenhorst, was also terrific to work with. Two anonymous reviewers provided helpful commentary and suggestions for improvement. They and the Princeton University Press Editorial Board have my appreciative thanks. Thanks to Anne Knight and Michael Blasgen, who gave me additional critical feedback on the book.

This book, by its very structure, could not exist without the support of MathWorks in Natick, Massachusetts, and its generous granting of permission to use its wonderful MATLAB as part of The MathWorks Book Program. I offer unbounded thanks for that support.

This book has a strong historical flavor, and my search for details on the early days of high-speed electronic computing greatly benefited from the following: correspondence with Laura Mullane, Nicholas Lewis, and Julie Miller at the Los Alamos National Laboratory, and with Sandra Fye at The National Museum of Nuclear Science & History, all in New Mexico, as well as with Heidi Polombo at the US Department of Energy in Washington, DC. The historical staffs at the Institute for Advanced Study (IAS) and the Niels Bohr Library & Archives helped locate several of the photographs in this book and determined their copyright status. Very special thanks in particular to Caitlin Rizzo (IAS Archivist) for permission to use the photo of John von Neumann standing next to the IAS Electronic Computer Project machine.

Back at Princeton University Press, when the book was "finally" written, it entered a new phase of oversight, and I thank production editor Kathleen Cioffi, copyeditor Nancy Marcello, and illustration wizard Dimitri Karetnikov for making that last push a smooth one.

This book was written over a period of two years in my favorite writing spot: an old, beat-up leather chair in a quiet corner of *Me & Ollie's Café* (the coffee drinker's *Cheers*—"where everybody knows your name"), on Water Street in Exeter, New Hampshire.

Finally, *as always,* I thank my loving wife Patricia Ann *for everything.*

Paul J. Nahin
June 2025

NOTES

Prelude

1. J. E. McKenna, "Computers and Experimentation in Mathematics," *The American Mathematical Monthly*, March 1972, pp. 294–295.

2. Despite the fairy tale allusion, the problems we consider in this book *are* (for the most part) serious.

3. Another type of such books is the 1981 *How To Solve It by Computer* (Prentice Hall) by R. G. Dromey, who was inspired by the classic book *How to Solve It* (Princeton, 1945) by Hungarian-American mathematician George Pólya (1887–1985). Dromey's book is a collection of detailed presentations on how to implement certain microtasks that commonly occur when writing computer programs for scientific computation (how to compute factorials, how to compute square roots, how to find the largest [smallest] of a set of numbers, how to find the greatest common divisor of two positive integers, how to sort numbers in increasing [decreasing] order, and so on).

4. The ability to do calculations so quickly (combined with a high-speed, solid-state memory—no moving parts!—of terabyte size) allows the PS5 to render live-action scenes such as windblown bodies of water so realistically that you need to look twice to convince yourself that the display screen is not actually wet.

5. The four-color map theorem, conjectured in 1852 by mathematician Francis Guthrie (1831–1899), states that any planar map, no matter how complicated, can always be colored, with at most four colors, so that countries with a common border have different colors. To this day, there is no known purely analytical proof. For a good discussion of the theorem and its computer proof, see Robin Wilson, *Four Colors Suffice: How the Map Problem Was Solved*, Princeton 2013.

6. One curious form of failure occurred when some foreign object (like a moth) landed on a relay contact and was then squished between the contacts. The fouled contact resulted in a disrupted circuit path, which meant an erroneous computation. The hunting down of the offending dried corpse and removing it got the obvious name of *debugging*, a term that survives today to describe trying to get a balky piece of computer code to run correctly.

7. You can read more about MANIAC and Fermi's interest in it in the essay by Herbert L. Anderson, "Metropolis, Monte Carlo, and the MANIAC," *Los Alamos Science*, Fall 1986, pp. 96–107. Physicist Nicholas Metropolis (1915–1999)—see Figure 1.2.1—was the leader of the team that designed MANIAC. I will say more about Monte Carlo later in the book.

8. For more on Kelvin's computer, see my *Dr. Euler's Fabulous Formula*, Princeton 2011, p. 170.

9. P.A.M. Dirac, "Quantum Mechanics of Many-Electron Systems," *Proceedings of the Royal Society of London A*, April 6, 1929, pp. 714–733.

10. Serber's once top-secret lectures are available today as *The Los Alamos Primer*, as a free pdf document on the Web. The *Primer* was declassified in 1965: during the war, revealing its contents would have resulted in a prison sentence, but today it's "just" freshman physics.

11. You can find a detailed description of the machine language instruction set for the IAS machine in Gerald Estrin, "The Electronic Computer at the Institute for Advanced Study," *Mathematical Tables and Other Aids to Computation*, April 1953, pp. 108–114. After reading that paper, I think you will certainly agree that life for people who write computer programs has greatly improved since the early 1950s!

12. As a little play on Kelvin's "brass for brains" remark, the message of this book might be labeled as that of substituting "sand for smarts." (High-speed solid-state electronics is based on silicon, of which there are megatons locked up in the sands of every beach and desert.)

13. See my *The Mathematical Radio*, Princeton 2024, for more on Hardy and his famously eccentric opinions concerning applied mathematics.

14. N. Metropolis, "The Beginning of the Monte Carlo Method," *Los Alamos Science*, Special Issue 1987, pp. 125–130.

15. Quoted from George Dyson, *Turing's Cathedral*, Pantheon 2012, p. 153.

16. See D. R. Hartree, "The ENIAC: An Electronic Computing Machine," *Nature*, October 12, 1946, pp. 500–506.

17. A fictional treatment of this dread is in the 1950s short story "God and the Machine" by English psychologist Nigel Balchin (1908–1970) reprinted in Clifton Fadiman's 1958 *Fantasia Mathematica*. This tale, one very different from those of Brown and Clarke with their omnipotent computers, has a checkers-playing machine designed by a physicist who admires the cold, hard logic of his machine as much as he despises the illogic of humans. He is, therefore, shocked to the core when he observes his machine, when confronted by a human opponent who plays a wildly unconventional game, make an illegal move. This story poses a question that perhaps *should* be taken seriously as artificial intelligence continues to be developed: which is more terrifying, a supersmart machine that never makes a mistake or a reasonably smart machine that cheats now and then?

18. Bernstein's essay is reprinted in his book *A Comprehensible World: On Modern Science and Its Origins*, Random House 1967, p. 191.

19. Adam Kirsch, "The Smartest Man Who Ever Lived," *The Atlantic*, November 2023, pp. 87–89, a review of Benjamin Labatut's *The MANIAC (Penguin Press 2023)*. As a literary work of remote (in time) psychoanalysis of a long-dead person, the book is noteworthy. As an accurate historical chronicle, however, read it with some skepticism. As two examples of what I mean by that, Labatut incorrectly names von Neumann's IAS machine as MANIAC, and completely ignores Metropolis (an omission particularly ironic since it was Metropolis who gave the *Los Alamos machine* the MANIAC name! See N. Metropolis and J. Worlton, "A Trilogy on Errors in the History of Computing," *Annals in the History of Computing*, January 1980, pp. 49–59 [particularly p. 56]). Labatut's error is not an uncommon one. Several years after the Metropolis/Worlton paper, for example, an eminent Harvard mathematical physicist (who shall remain nameless here), in a scholarly paper in a well-known math journal, claimed that ENIAC was the Princeton computer. Well, of course, . . . no.

20. Michael Blasgen, who, after a long career in computer architecture development at IBM, capped it off as Vice President for Research at Sony Electronics.

Chapter 1. Computers That Play Games

1. You can read about the bomb work in the fascinating book by Kenneth W. Ford, *Building the H-Bomb*, World Scientific 2015. Ford (born 1926) was a graduate student in the Princeton physics department recruited to handle the computer studies of various fusion bomb geometries. Ford worked at both Los Alamos and Princeton from 1950 to 1952, but all his work was done using earlier computers, and not on either MANIAC-I or the IAS machine, both of which became operational only months before the first H-bomb detonation (the 10 megaton *Ivy Mike*) in November 1952.

2. A more extensive discussion of one of them, the famous Fermi-Pasta-Ulam-Tsingou experiment, is in my book *Number-Crunching*, Princeton 2011, pp. 55–73. Fermi thought the result of that particular MANIAC-I computer experiment to be a significant discovery in atomic physics. See also N. Metropolis, "The Los Alamos Experience, 1943–1954," *A History of Scientific Computing*, ACM Press 1990, pp. 237–250.

3. Claude Shannon, "Programming a Computer to Play Chess," *Philosophical Magazine*, March 1950 (written in 1949), pp. 256–275. Shannon is famous for introducing Boolean algebra as a now standard tool for designing digital circuitry, and for almost single-handedly "inventing" information theory. A dual biography of Shannon and the English mathematician George Boole (1815–1864) is in my book *The Logician and the Engineer*, Princeton 2017.

4. Jonathan Schaeffer *et al.*, "Checkers Is Solved," *Science*, September 14, 2007, pp. 1518–1522. See also Schaeffer's book *One Jump Ahead*, Springer 2008. Schaeffer (born 1957) is a professor of computer science at the University of Alberta.

5. This is not to say checkers is trivial. Far from it! A poetic way to distinguish the two games is due to Marion Tinsley (1927–1995), generally thought to be the best human player of all time (he was world champion numerous years, as well as a university professor of mathematics): "Chess is like looking out over a vast ocean. Checkers is like looking into a bottomless well."

6. It is not clear to me how Shannon arrived at that number. There are about 30 million seconds $= 3\times10^7$ in a year, and so our assumed speedy computer could run through $3\times10^7\times10^6 = 3\times10^{13}$ chains (games) in a year. To run through 10^{120} games would require $\frac{10^{120}}{3\times10^{13}} = \frac{10\times10^{119}}{3\times10^{13}} \approx 3\times10^{106}$ years, which is *much* larger than Shannon's number (making, of course, his point even more dramatically).

7. John von Neumann and Oskar Morgenstern, *Theory of Games and Economic Behavior*, first published in 1944 by Princeton University Press (and never out of print).

8. This possibility would be a real surprise to most chess players as it is generally assumed that to go first should not be a disadvantage. But, since the optimal **S** is unknown for chess, the possibility of Black being able to force a win from the get-go can't be ignored. Shannon does mention, in passing, one strategy that is *far* from optimal (but easy to program). He writes, "It is possible for the machine to play legal chess [by] merely making a randomly chosen legal move at each turn. . . ." But, as he concludes, "The level of play with such a strategy is unbelievably bad." He estimates (by some unexplained calculation) that the probability such a strategy would win a game against a world champion is of the order of 10^{-75} (see Shannon's paper in note 3).

9. P. Stein and S. Ulam, "Experiments in Chess on Electronic Computing Machines," *Chess Review*, January 1957, pp. 13–17. A few months later a variant of that paper appeared under the title "Experiments in Chess," *Journal of the Association for Computing Machinery*, April 1957, pp. 174–177.

10. Two heuristic measures of "merit" were used to select a move: (1) the *strength* of the resulting chessboard (using commonly accepted relative values of the individual pieces, starting with the weakest, the pawn) and (2) the *mobility* of the resulting chessboard (the number of legal moves available after the proposed move (the more legal moves available, the better). MANIAC made its selection of a move based on maximizing some combination of these two measures.

11. A match is scored as follows. A win gives 1 point to the winner and 0 points to the loser. A drawn game gives $\frac{1}{2}$ point to each player. A match consists of an *even* number of games, so that each player has the White pieces (and so moves first) an equal number of times. In 1997 the IBM chess computer *Deep Blue* defeated the

then–world champion Gary Kasparov in a six-game match (Kasparov won one game, *Deep Blue* won two games, and the other three games were draws) with a final score of $3\frac{1}{2}$ to $2\frac{1}{2}$. It is estimated that Kasparov and *Deep Blue* were pretty evenly matched, with each having a rating of about 2,900. *Stockfish* would almost certainly beat them both today. But that would be of no impact because, in 2017, a new computer code named *AlphaZero* defeated *Stockfish* in a 1,000 game match with a performance of 155 wins, 839 draws, and just 6 losses. *AlphaZero* has a rating of perhaps 3,800, a *thousand* points beyond that of the human world champion.

12. J. Marshall Ash, "The Probability of a Tie in an *n*-Game Match," *The American Mathematical Monthly*, November 1998, pp. 844–846.

13. MATLAB has a large number of special mathematical functions that make the job of coding *immeasurably* easier than it was for yesteryear analysts who had to do their work in machine language. In particular, the MATLAB command *nchoosek* computes binomial coefficients: if *top* and *bottom* are two given nonnegative integers, then *nchoosek(top,bottom)* computes $\binom{top}{bottom}$. To program this in machine language would be a (*very*) nontrivial task.

14. Named after Scottish mathematician James Stirling (1692–1770). You can find a high school level (if you have had AP-calculus) derivation and discussion of the approximation in my book *The Probability Integral*, Springer 2023, pp. 136–144. Stirling's formula is said to be *asymptotic* because it is *not* an equality but instead is an approximation that becomes better (decreasing *relative* error) as *m* increases. The formula is remarkably accurate: by the time *m* reaches 8, the error is less than 1 percent.

Chapter 2. A Gambling Problem (the Monte Carlo Concept)

1. One math historian has made the important point that it may not be appropriate to call Buffon's needle tossing experiment a *simulation* (which means an "attempt to learn about a process by analogy"). As Stephen Stigler continues in his paper "Stochastic Simulation in the Nineteenth Century," *Statistical Science*, February 1991, pp. 89–97, "If the experimental value of π had come out far from 3.1416 then the experiment would have been discarded, not the "'*mathematically derived* value of π.'" To put it bluntly, nobody does the Buffon experiment with the hope of learning anything new about π!

2. So named because it appears in Euclid's work, showing he knew lots more math than just plane geometry! We really don't have to use the algorithm to answer our question, of course; it is mostly a matter of taste. You can find the algorithm in any good book on elementary number theory, or in my book *How to Fall Slower than Gravity*, Princeton 2018, pp. 137–146, which gives a very simple, very fast MATLAB program that implements the algorithm.

3. What does "setting the odds" mean? Here's one way to look at it. Since the probability of no match is 0.56, then the probability of a player getting a match is 0.44. Suppose the casino says the player wins a game if a match occurs, and sets the charge to play at $5 a game. If the player loses she is out her money, but if she wins she gets her money back plus another $5. In 1,000 games the casino "expects" players to win 440 times, resulting in the casino paying out $4,400. This is fine with the casino because it will take in $5,000, resulting in a profit of $600 per 1,000 games.

Chapter 3. The Carpenter's Problem

1. You can find the math details of this cute problem in the book by Howard W. Eves, *The Other Side of the Equation*, Prindle, Weber & Schmidt 1971, pp. 72–73.

2. When we use the word *randomly*, with no elaboration, we will always assume a *uniform* distribution (see the Appendix).

3. In **method1.m** the value of *l* is taken to be 1, which is okay to do since the *probability* of a triangular frame is independent of *l*. In our analysis, you will have noticed that *l* cancelled away). The value of *l* serves as a scaling factor that determines the *size* of a frame.

Chapter 4. Wi-Fi Coverage and Antisubmarine Warfare Are the Same Math Problem

1. It is easy to generate points at random over a square, as you will see in just a moment. If we accept only those points that are also inside the restaurant circle, then those accepted points are obviously random over the interior of the restaurant circle, too. There is a way to *directly* generate points at random over a circle, without a rejection step, but the price paid for that is an increase in the computation required (see my book *Digital Dice*, Princeton 2013, pp. 16–18 for details).

2. This quotation is repeated in L. A. Graham, *Ingenious Mathematical Problems and Methods*, Dover 1959, which then states the so-called *Mrs. Miniver problem*: If the two circles are generally not the same size, when is Mrs. Miniver's suggested condition satisfied? I won't pursue this here, but you can find a solution on pp. 64–66 of Graham's book.

3. William C. Guenther, "On the Probability of Capturing a Randomly Selected Point in Three Dimensions," *SIAM Review*, July 1961, pp. 247–251. Professor Guenther celebrated his hundredth birthday on December 21, 2021.

4. My favorite because, in the late 1960s and early 1970s (before I became a professor), I worked as a weapons systems analyst, first at Hughes Aircraft (in Southern California) and then two Federal Contract Research Centers in Washington, DC (The Institute for Defense Analyses [IDA], and The Center for Naval Analyses [CNA]). The problem treated by Professor Guenther is just like many of those I was

tasked at IDA and CNA to think about, and so reading his paper was a trip back in time for me. Great fun!

5. Professor Guenther implicitly assumes the torpedo is a dumb weapon with no active tracking capability. The torpedo simply does its best to go to where the hostile submarine was *when first detected*. A modern torpedo, of course, like the US Navy's Mk-48, has the capability to track the target's instantaneous position and to adjust its own trajectory during the travel time to the target. Warhead detonation is initiated via proximity detection, not simply by arriving at some predetermined location.

6. As discussed in the Appendix, the probability density function of a zero-mean, normal random variable with standard deviation σ is $\frac{1}{\sqrt{2\pi}}e^{-x^2/2\sigma^2}=\frac{1}{\sqrt{2\pi}}e^{-\frac{\left(\frac{x}{\sigma}\right)^2}{2}}$. So, if we measure x in units of σ, then the pdf of the aiming error in spatial direction x is $\frac{1}{\sqrt{2\pi}}e^{-x^2/2}$.

7. Notice that the P_{kill} expression does make physical sense for the special case of $R=0$. That is, since $R=0$ means the torpedo warhead has a lethal radius of zero (and so is not of much military interest!), then P_{kill} should be zero, which is just what Professor Guenther's expression clearly reduces to when $R=0$.

8. See Chuck Hansen, *U.S. Nuclear Weapons: The Secret History*, Orion Books 1988. If we move up in warhead size, R increases even more dramatically. In his book *Arsenal: Understanding Weapons in the Nuclear Age (Simon and Schuster 1983)*, Kosta Tsipis writes, "a 1-megaton weapon has on average a lethal radius against a submarine of about 6,000 meters" (p. 233).

Chapter 5. A Problem in Electric Circuit Theory

1. For the literal minded, there is no such thing as a one-volt *battery*. Battery voltages are determined by the inherent physical properties of the chemicals used in the battery. An ordinary dry-cell flashlight battery (which isn't actually dry!) is about 1.5 volts, a value determined by chemical physics, not by specific design. Saying we have a one-volt battery here is simply a convenient shorthand for a one-volt *source* of some design.

2. The use of *volt* is so common that it seems almost pedantic to elaborate but, just to be complete, the unit is named after Italian physicist Alessandro Volta (1745–1827). For a sense of scale, a charged car battery is about 12 volts, and a high-power long-distance transmission line is typically many hundreds of thousands of volts.

3. *Negatively* charged electrons actually move from – to + (opposite charges attract) in Figure 5.1.1, which we think of as a *positive* current flowing from + to –. This sign issue is, in my opinion, the cause of 90 percent of the difficulty students have when first studying this material.

4. The unit of current, the ampere, is named after French physicist André-Marie Ampère (1775–1836). To give you a sense of scale, a typical household circuit breaker is something like 15 or 20 amperes, and the current a car battery provides to an electric motor to turn over an engine on a cold morning is typically 200 to 400 amperes.

5. There is nothing that says we must label the nodes as is done in Figure 5.2.1, but in fact I *did* follow a simple recipe. Think of the node labels as binary numbers, that is as running from 000 to 111. Then notice that each node connects to the three nodes with labels that differ in just one position. For example, 000 connects to 001, 010, and 100. That is, node ⓪ connects to nodes ①, ②, and ④). And 001 (node ①) connects to 000, 011, and 101 (nodes ⓪, ③, and ⑤). And this continues for all the other nodes. When we start to develop the computer algorithm that solves any resistor circuit, we will follow a different recipe for labeling nodes.

6. I will return to the issue of the battery current at the end of this section. But, for now, forget about it.

7. Matrix multiplication computes the term-by-term sum of the products of the selements of a *row* in $\boldsymbol{A}$ with the elements in the *column* vector $\boldsymbol{v}$.

8. Here's the simple MATLAB code that calculates $\boldsymbol{v}$. The three dots at the end of the first line defining $\boldsymbol{A}$ is MATLAB stating the definition is continued on the next line):

```
%lattice.m
b=[-4;-2;0;0;0;0];
A=[-8 0 2 1 0 0;0-9 1 4 0 0;4 1-9 0 2 2; . . .
4 8 0-29 1 16;0 0 4 1-9 4;0 0 1 4 1-6];
v=inv(A)*b
```

9. By now you are almost certainly wondering why I have always specified the battery voltage in the circuits we have studied as being one volt. What is so special about *one* volt? As we develop the computer code that I have mentioned so often, you will see an intimate connection between the battery voltage and the *probability* value of 1.

10. I don't know the historical origin of the approach I am about to show you, but the intimate connection between random walks and battery-energized resistor circuits has been known for a long time (at least it has among mathematicians but, perhaps, not so much among electrical engineers). See, for example, an elegant little book (which, alas, is often at a math level just above that of this book) by Peter G. Doyle and J. Laurie Snell, *Random Walks and Electrical Networks*, Mathematical Association of America 1984.

11. The transition probability rule of (5.5.5) *does* have a nice *physical* motivation that appeals to physicists and electrical engineers. Since $p_{x\to y}=\dfrac{1/r_{xy}}{\sum_{y\in N(x)}\frac{1}{r_{xy}}}$, then

when an electron is at node x and faced with "deciding" among multiple possible paths to a connected node, the rule gives candidate nodes that present *higher* resistance paths and *smaller* probabilities of being chosen. That is, the electron tends to pick the "path of least resistance." And isn't that what *you* would do, too, if you were a random walker?

12. There *is* the physical possibility of negative resistance, but only in *electronic* circuits (those with vacuum tubes and/or transistors), not in strictly all-resistor circuits.

13. Lines 16 and 17 define the vector V, the elements of which will eventually be the node voltages of the circuit. The reason $V(1)$ is assigned the value of ten million is that the code will divide all the elements of V by ten million to arrive at the node voltages (Line 39) and, of course, $V(1) = 1$. This division occurs because, before the division, the element $V(i)$ is the number of walks (out of ten million), starting at node i, that terminate on node 1.

14. Rayleigh's law appears in the famous 1873 *A Treatise on Electricity and Magnetism* (Clarendon Press) by James Clerk Maxwell (1831–1879), a work often ranked with Newton's *Principia* in importance in the history of science. In it, Maxwell stated the law to be "self-evident." Well, maybe so, but you can find a more analytical and extensive discussion of the law in the book by Doyle and Snell (note 10).

Chapter 6. Simulation of a Purely Deterministic Problem

1. Turing did return to America in a 1943 wartime visit to the Bell Telephone Laboratories (where Shannon worked). The two men had a significant interaction during that visit, which influenced Shannon's thinking about what eventually led to his development of information theory.

2. You can read more about vacuum-tube triodes and their operation in Chapter 3 of my book *The Mathematical Radio*, Princeton 2024.

3. If you *are* interested in the details of designing a clocked, sequential, finite-state, synchronous digital machine, take a look at my *The Logician and the Engineer*, Princeton 2017 (particularly Chapter 8).

4. Shannon argued that one measure of the complexity of a Turing machine is the product of the number of states in its finite-state machine component, and the number of different symbols the Turing machine can recognize. In other words, for a given complexity the number of states and the number of symbols can be traded off against each other. Shannon took that trade-off to the limit in both directions: for a given complexity, one can have just two states if there are enough symbols, or one can have just two symbols (as we are assuming) if there are enough states.

5. From Minsky's beautiful book *Computation: Finite and Infinite Machines*, Prentice Hall 1967, p. 128.

6. Have you noticed the behavior of this machine when computing 0 + 0? What would be the easiest calculation of all to do?

7. Tibor Radó, "On Non-Computable Functions," *Bell System Technical Journal*, May 1962, pp. 877–884.

8. Shen Lin and Tibor Radó, "Computer Studies of Turing Machines," *Journal of the Association for Computing Machinery*, April 1965, pp. 196–212. You will recall (from the Prelude) mention of Appel and Haken's use of computers to solve the four-color map problem. That use of a computer is the one typically cited as the first such use to solve a math problem. In fact, however, the Lin-Radó computer proof predates the Appel-Haken proof by more than a decade.

9. For $n = 4$ the number of candidate machines is 20^8, or more than 25 billion, a *huge* increase over the $n = 3$ case. Brady's conjecture appeared in the *IEEE Transactions on Electronic Computers*, October 1966, pp. 802–803.

10. Allen H. Brady, "The Determination of the Value of Radó's Noncomputable Function $\Sigma(k)$ or Four-State Turing Machines," *Mathematics of Computation*, April 1983, pp. 647–665.

11. Marxen Heiner and Jùgen Buntrock, "Attacking the Busy Beaver 5," *Bulletin of the European Association for Theoretical Computer Science*, February 1990, pp. 247–251.

12. One place to look for the story is Clifton Fadiman's wonderful 1958 collection of "mathematical literature," *Fantasia Mathematica* (Springer-Verlag).

13. Porges cleverly expanded on the dangers of negotiating with the Devil in a later story, "A Devil of a Day." If you can find a copy of the August 1962 issue of *Fantastic Magazine*, you will see that the Devil is, indeed, a tricky fellow. The original pulp printing has long since oxidized to nothingness, so try the microfilm archives at a college library.

Chapter 7. Schwartz's Question: The Toughest Problem in This Book

1. Marilyn vos Savant, "Ask Marilyn," *Parade Magazine*, May 2, 2010, pp. 24–26.

2. When I wrote this, I was assuming the walking and riding speeds were the same for the two boys, an assumption I used in the published *Number-Crunching*. Here, in this book, we will generalize the problem just a bit by allowing the two speeds of walking and riding to be independently assigned for each boy (with the caveat that each one rides faster than the other one walks—otherwise leapfrogging simply doesn't work!).

3. Richard Feynman, *Surely You're Joking, Mr. Feynman!*, W. W. Norton 1995, pp. 250–251.

4. Schmidt eventually published his solution as the paper "Two Friends and a Bike," *The College Mathematics Journal*, January 2021, pp. 11–21. Nerinckx did not formally publish, but instead wrote it all up in a little essay he sent me as a pdf, along with permission to share it with anyone who asked (write to me at paul.nahin@unh.edu for a copy).

5. At the heart of the code is the time variable *t*, initially set equal to zero, which is then incremented in steps of 0.001 second. When the locations of the boys, and of the bike, are updated, their motions are how far they have moved (at their current speeds) in 0.001 second. At a speed of 6 mph, for example, that distance is 0.0088 feet, which is just a bit more than one tenth of an inch.

6. If you think you can skip doing a flowchart for leapfrogging, well, good for you. But, not to sound (too much) like a grumpy old grandfather, there will *always* be problems—for *everyone*—that are sufficiently complicated that you will find yourself going into "nutso land" if you start coding without a flowchart to guide you. So, even if you think it unnecessary for leapfrogging, consider Figure 7.2.1 as good practice for all those *really* convoluted problems you will surely encounter in the future.

Chapter 8. A Counterexample Concerning Computer Simulations

1. Random walks occur in a multitude of real-life problems, and I discuss some of them in my book *Mrs. Perkins's Electric Quilt* (Princeton 2009). See, in particular, pp. 233–284.

2. With p and q defined in the text, the drift is clearly to the right. If we instead wrote $p=\frac{1}{2}-\epsilon$ and $q=\frac{1}{2}+\varepsilon$, the drift would be to the left. For what we are doing in this chapter, the choice is arbitrary and our conclusions will hold for either choice.

3. William Feller, *An Introduction to Probability Theory and Its Applications*, John Wiley & Sons 1968 (3rd ed.), pp. 86–88. Feller was professor of mathematics at Princeton and, at the time of his death, lived on (how appropriate!) Random Road. Feller had a sense of humor, as I found out when I wrote to him just before his death. His photograph had recently appeared on the cover of a technical magazine and I sent it to him with a request for an autograph. Back came a fast reply with a note saying he wondered if I had perhaps confused him with the great Cleveland Indians baseball pitcher Bob Feller—but, on the chance I wasn't confused, he had gone ahead and signed it!

4. For a funny spoof of the "law of averages," and of what could happen if it "fails," see the essay "The Law" by Robert M. Coates. Originally published in 1947 by *The New Yorker magazine*, it was reprinted in Clifton Fadiman's 1962 *The Mathematical Magpie* (a sequel to his 1958 *Fantasia Mathematica*).

5. Pólya showed an eternal return to the starting point for a symmetrical walk holds not only in the one-dimensional case but also in the two-dimensional case. In *three* dimensions, however, Pólya further showed that even for a symmetrical walk there is a nonzero probability (slightly greater than 0.34) of the walker escaping to infinity. This explains the drunk bird joke that opened this chapter. Here is another version of that joke that I particularly like: "It's better, if you want to find your way back home, to get drunk in [flat] Nebraska than it is to get drunk in [high-rise] Manhattan." Who says mathematicians have no sense of humor!

6. A simulation can, however, *hint* at the discontinuity of the return question as epsilon goes from zero to nonzero. I will return to this point later in the chapter.

Chapter 9. Monte Carlo Integration

1. Francis Y. Kuo and Ian H. Sloan, "Lifting the Curse of Dimensionality," *Notices of the AMS* (American Mathematical Society), December 2005, pp. 1320–1328.

2. It can be shown that, under very general assumptions, the error made by a Monte Carlo integration, using n darts, decreases as $\frac{1}{\sqrt{n}}$. That is, if we wish to decrease the error by a factor of 10, then n must be increased by a factor of 100.

3. If you are curious about contour integration, see my book *Inside Interesting Integrals*, Springer 2020 (2nd ed.), pp. 351–422 (Chapter 8). The discussion there starts with the assumption you have an AP-level background in calculus (and know what a complex *number* is), but complex *functions* need to be explained.

Chapter 10. Gamma-Ray Path across a Semicircle

1. W. Primak, "Gamma-Ray Dosage in Inhomogeneous Nuclear Reactors," *Journal of Applied Physics*, January 1956, pp, 54–62. Gamma-rays are *extremely* high-frequency electromagnetic radiation (10^{20} hertz and higher—by comparison, a typical AM broadcast radio station frequency is 10^6 hertz) that are deadly to living organisms. Gamma-rays are used, for example, to sterilize food.

2. As a quadratic equation, (10.2.2) will give us two values for x_1, and we will use a physical argument to resolve the ambiguity of the $\pm$ sign in (10.2.3).

3. Perhaps even more physical is the observation that if we used the minus sign, we would have $l_1 < 0$, which makes no physical sense.

4. For a different geometry, a new path analysis must be done. For the case of random paths across a square, for example, see my book *Duelling Idiots and Other Probability Puzzlers*, Princeton 2002, pp. 58–61, 147–155, and 244–245, for both Monte Carlo and theoretical treatments.

5. From Damon Knight's short story "Extempore," first published in the August 1956 issue of *Infinity Science Fiction Magazine*.

Appendix: Got a Math Question? Try Looking Here

1. N. Metropolis, "The Age of Computing: A Personal Memoir," *Daedalus*, Winter 1992, pp. 119–130. Metropolis wrote this decades after his leadership role in building the MANIAC-I high speed, electronic (vacuum-tube) computer, and he was then a Senior Fellow Emeritus at Los Alamos National Laboratory.

2. In his memoir, Metropolis makes extensive remarks on how American physicist Richard Feynman (1918–1988) became interested in reliability theory while at Los Alamos in his efforts to get useful performance out of the pre-electronic electromechanical machines he had to use in his computational work.

3. For lots more about de Moivre the man, and his contributions to the development of probability theory from faulty intuition to an analytical basis, see my book *The Probability Integral*, Springer 2023.

4. By "Chance of a Prize" de Moivre meant for there to be a probability of at least half of winning a prize. As you have no doubt assumed, a blank is a nonprizewinning ticket.

5. Writing this statement is an example of what I earlier called *casual*. I am taking advantage of the "common knowledge" that zero is the probability of an impossible event, and 1 is the probability of a certain event. A mathematician could write a *book* on the formal establishment of those two assumptions, but I will just take both for granted. In a similar fashion, I will take the idea of a negative probability as being without meaning.

6. Francis Galton, "Dice for Statistical Experiments," *Nature*, May 1, 1890, pp. 13–14.

7. Kelvin's approach to account for the inescapable bias in any real coin was *not* the method of double flipping discussed in section A.2. As Kelvin explained, the interpretation of heads and tails was "frequently changed [that is, reversed] to avoid the possibility of error by bias." What *frequently* meant was not explained. Kelvin also discussed another approach for random number generation but, in the end, rejected it: "I had tried numbered small squares of paper drawn from a bowl but the best mixing we could make in the bowl seemed to be quite insufficient to secure equal chances. . . . In using one's finger to mix [small squares of paper] in a bowl, very considerable disturbances may be expected from electrification [static electricity]."

8. John von Neumann, "Various Techniques Used in Connection with Random Digits," *Journal of the National Bureau of Standards*, Series 3, 1951, pp. 36–38. This essay was actually written by American mathematician and computer scientist George Forsythe (1917–1972), as a summary of a talk given by von Neumann.

9. In his 1991 book *The Broken Dice and Other Mathematical Tales of Chance*, French mathematician Ivar Ekeland tells a curious story about how the middle-square method may have been discovered centuries before von Neumann. He writes that Argentine fantasy writer Jorge Luis Borges (1899–1986) sent him a copy of a

manuscript that has, alas, "unfortunately disappeared" from the Vatican archives where Borges claimed to have found it. This manuscript, dating from the mid-thirteenth century, was written by a Norwegian monk (one Brother Edvin) of the Franciscan Order. In this amazing document Brother Edvin describes the middle-square method! Ekeland seems to take this story (which strikes me as a possible plot for a sequel to *The Da Vinci Code*) quite seriously, but I think math and computer science historians continue to strongly lean toward crediting von Neumann.

10. One signature of a "bad" random number generator is a short period, and 10,000 is laughably *far* too short for use by modern Monte Carlo codes that consume random numbers by the *tens of millions*.

11. See N. Metropolis et al., "Equation of State Calculations by Fast Computing Machines," *The Journal of Chemical Physics*, June 1953, pp. 1081–1092. The middle-square generator used in this paper had a period of about 750,000. (An equation of state relates the pressure, volume, and temperature for a given mass of matter. Every high school chemistry student has encountered such an equation when studying the ideal gas law.)

12. The notation (n) *mod* N means "divide n by N and keep the remainder." For example, (7) *mod* $3 = 1$ and (7) *mod* $7 = 0$.

13. In one version of MATLAB (now years old), the Twister period was 2^{1492} (for the obvious reason, this generator was called the *Christopher Columbus generator*). Its period is so large that if it had started to generate random numbers at the instant of the Big Bang, at the rate of one trillion per nanosecond, then as of today it would have run through only an infinitesimal fraction of the period. Newer versions of the Twister have periods that are immensely greater than that of the Christopher Columbus.

14. See Roger Eckhardt, "Stan Ulam, John von Neumann, and the Monte Carlo Method," *Los Alamos Science*, Special Issue 1987, pp. 131–143. A photo of von Neumann's letter is reproduced on p. 135 of that paper.

15. In the essay cited in note 8, von Neumann reminds his audience that this pdf occurs in the study of "free paths [of neutrons]" moving through matter and suffering collisions with the molecules of that matter (think of a chain reaction in an atomic bomb warhead). You can find a freshman calculus derivation of this result in Chapter 43 of Volume 1 of Feynman's famous *Lectures on Physics* (1964) (all three volumes of the *Lectures* can be read on the Web).

16. For a proof of this, see my book *The Probability Integral*, Springer 2023, pp. 73–79.

17. The first such computer calculation was done in 1950 on the ENIAC, which calculated the first 2,037 decimal digits of pi.

18. An interesting historical discussion of the randomness of pi, and how modern computer calculations of pi are done, is in a paper by Yadolah Dodge, "A Natural

Random Number Generator," *International Statistical Review*, December 1996, pp. 329–344.

19. I say more about the mathematics of Sagan's presentation of "The Artist's Signature" in my book *Holy Sci-Fi!: Where Science Fiction and Religion Intersect*, Springer 2014.

20. You can find further discussion of this amazing continued fraction, and Euler's derivation of it, in my book *How to Fall Slower than Gravity*, Princeton 2018, pp. 243–244, 247–250.

21. S. M. Ulam, *Adventures of a Mathematician*, Charles Scribner's Sons 1976, pp. 196–200. See also Nicholas Metropolis and S. Ulam, "The Monte Carlo Method," *Journal of the American Statistical Association*, September 1949, pp. 335–341.

INDEX